Techniques and Fieldwork in Geography

Collins Educational
An imprint of HarperCollins*Publishers*

Published by
CollinsEducational
77–85 Fulham Palace Road
London W6 8JB
An imprint of HarperCollins*Publishers*

First published in 1983 by University Tutorial Press Limited
Reprinted 1984, 1986, 1987, 1988, 1990
Reprinted 1992 by CollinsEducational

ISBN 0 00 322277 2

Printed in Great Britain by Scotprint Ltd, Musselburgh

Contents

Preface

In this book we have tried to do three things. Firstly, to produce a comprehensive text of a wide range of techniques of data measurement and analysis needed in the Sixth Form. Secondly, to present these in a form which can readily be understood by students with little or no previous experience of such techniques. For this reason, methods which are particularly complex, as well as those involving expensive pieces of apparatus, have been excluded. Thirdly, to give a number of examples of possible fieldwork projects which may act as a guide or inspiration.

The division of labour for writing the book was thus: Paul Cleves wrote the chapter on physical geography; Barnaby Lenon was responsible for all the remaining chapters in the book.

We would particularly like to acknowledge the help given by David Collins, Michael Town, Pauline Hanson and Janet Easterling.

Barnaby Lenon and Paul Cleves
Eton, 1990

Introduction

The most important change in the nature of geographical study over the last ten years has been the growth of practical and applied work. This is a reflection of both developments in the subject matter and the emergence of a wide range of useful techniques. Some of these techniques, such as network analysis, are largely intellectual. Others are much more practical, for instance the use of instruments developed for measuring stream discharge or soil infiltration rates.

These changes have been reflected in A-level and university courses. Many books have been written about them but most are fairly complicated and difficult to read. This book aims to overcome the problem by being both comprehensive and easy to understand. Every technique is described in a step by step fashion, using straightforward language, diagrams and worked examples. All the methods described can be done by students with limited time and resources and the inclusion of many suggestions for project work is a reflection of this. It can either be read as a textbook or taken out into the field and used as a practical work manual.

1 Fieldwork projects

1.1 How to choose a project title

The most difficult but important part of a project is deciding at the outset exactly what to study. A lot of thought and preparation at this stage will save time and frustration at later stages when you are trying to collect information for your project. It is important to choose a subject which interests and motivates you. However, your subject *must* be geographically appropriate. Equally important it must be feasible in terms of the time, facilities and equipment available to you. It is no good planning a study of residential patterns in North American cities if you cannot visit them yourself and therefore have to rely entirely on other people's research.

A number of strategies which you can employ to help you make a wise decision on project title are outlined here:

1 **Bear in mind the fact that three types of project tend to be particularly successful:**

a Those which test a hypothesis or theory, such as von Thünen's theory that the intensity of agricultural land use falls as one moves away from a market. Virtually all aspects of geography have theories attached to them and these can be read in any standard textbook. Alternatively you may like to devise a hypothesis of your own.

b Those which compare the geographical characteristics of two places or phenomena, such as two shopping centres or two streams. A variation on this theme is a comparison of the geographical characteristics of *one* place or phenomena at two or more stages of time, i.e. a study of changes over time.

c Those which study a problem or land use conflict such as the environmental impact of building a reservoir or the consequences of a housing redevelopment scheme.

2 Think through the complete range of possible topics Consulting your course syllabus may help you here. Alternatively, consider the following selection of topics:

Human geography

Settlement patterns; urban land uses and land values; shops; the Central Business District; office location; industrial location, linkage and change; the spheres of influence of settlements.

Social geography: locational patterns of friends and relatives, ethnic and social class groups; population changes over time; migration.

Rural land uses; the changing character of rural areas.

Perception of landscapes, distance, and where people would like to live.

Recreation: patterns and conflicts.

The Environment: noise, pollution; mining; motorways; reservoirs; conservation measures.

Transport: ports; social and economic impact of new links; traffic flows; gravity models.

Locational analysis of swimming pools, garages, golf courses, Roman villas, schools, telephone kiosks, etc.

Physical geography

Geology: influence of rock type and structure on landforms and land uses.

Hydrology: the hydrological cycle; precipitation; evapotranspiration; soil moisture; groundwater levels; surface runoff.

Rivers: discharge; cross profile; hydraulic radius; gradient; load; meanders; deposition; drainage patterns.

Soils: characteristics; changes up a slope; relation to crop yields or vegetation.

Vegetation: distribution of species related to soils, slopes, microclimates, drainage, and human factors.

Ice action: erosional and depositional landforms.

Coasts: landforms; changes over time; wave energy; longshore drift.

Meteorology: weather recording and interpretation.

Climate: climatic trends and fluctuations; microclimates of urban areas, woodlands, lakes and coasts.

3 Be led by certain additional considerations:

a What has particularly interested you in your course?

b Would any of your other subjects help (e.g. economics, chemistry)? For instance, if you are studying chemistry you may be interested in examining the chemical composition of different horizons in a soil profile.

c Have you any special interests or hobbies which might be relevant? If you are a keen angler, for example, you may wish to investigate stream pollution.

d Do you have any special contact or sources of information? You may have access to the records of a family business, for example, which might be usefully incorporated into a project on the changing location of markets for a particular commodity.

e Is anything particularly interesting happening in your local area? Topics such as controversial new road schemes, the siting of a proposed refuse dump, or the growth/decline of a local industry might be investigated in project work.

4 Choose a fairly narrow subject. It is often better to study one stream in detail rather than five streams superficially, one shopping street rather than a whole town. On the other hand, try to avoid looking at just one factory or farm unless it is very large or interesting.

5 Choose a subject which will enable you to collect most of the information yourself from primary sources rather than relying on someone else's data or publications. This is one reason for avoiding a straightforward study of a New Town – so much will already have been published by the town's Development Corporation.

6 Do choose a topic which is geographical, in other words which deals with locations, patterns as they change over an area, or the relationship between man and the environment. Descriptions of factory processes or the workings of a farm are *not* strictly geographical.

1.2 How to carry out a fieldwork project

1 Your project will be divided into three stages:
 a Collecting the data.
 b Summarising and analysing the data using maps, diagrams, statistics and methods of spatial analysis.
 c Describing stages (a) and (b) in words.

2 Ensure that you are organised and have thought through every possible problem before you begin to collect the data. This is particularly important with physical geography projects, which require the collection of equipment and preparation of tables on which to record results. Questionnaires, too, must be carefully planned and tested.

3 Be flexible because problems always arise – rivers dry up, people refuse to answer questionnaires, and other sources of data may simply fail to come to anything.

4 Remember to study *processes* as well as patterns. If, for example, you study land uses in a town or the distribution of given shop types, it is important to try and explain *why* these patterns have evolved. This will normally mean consulting the local history section of a library, the use of questionnaires and interviews, and some intelligent guesswork.

5 The length of your written account will normally be specified by your examination board. A popular and successful way of organising this account is to divide it into the following sections:
 a Aims
 b Methods used
 c Analysis and results
 d Conclusions
 e Appendices (of data)
 f Bibliography (books or articles you have used)

Diagrams must be integrated into the text, numbered and referred to in the written account. Long tables of data, questionnaires and large numbers of maps should be put in the appendices at the end.

Finally, here is a list of some project titles which have proved to be successful for A-level candidates in the past. Reading these may suggest ideas for similar projects:

A search for patterns in Guernsey's glasshouse industry.
Hang-gliding: the study of a growth sport and its impact on Mill Hill, Shoreham-by-Sea.
An analysis of settlement patterns in SE Devon.
An investigation into erosion at Cooden Beach.
Is there a positive correlation between rainfall intensity and surface runoff? How does the use of a model help to predict this runoff under varying angles of slope?
A comparison and contrast between two Highland sheep farms in the far north.
A determination of the ecological consequences of visitors to Hound Tor.
Factors influencing the spatial distribution and siting of Roman villas in North Oxfordshire.
A geological study of Marmande Escarpment.
A soil study of Gironde.
A study of rural land use in an area south-east of Guildford.
The development of Swindon and the possibility of changes in the population structure as a result.
A study of 'gentrification' in Sunderland.
A study of the agricultural land use around Ndola – a city in a developing country.
Petrol retailing: spatial aspects of pricing policy on a micro-scale.
Impact on local soils of hedge removal.
Spatial diffusion on a Cornish beach.
Problems of land drainage in Amberley, West Sussex.
The Doddington Estate – a failure in high density housing.
Dorchester: a study of the Central Business District.
A study of the retail geography of Salisbury city centre.
Solute and sediment rating curves for two Warwickshire rivers.
The calculation, comparison and attempted explanation of stalagmite growth rates in a Bermudan cave.
A study of types of building stone in relation to local geology in SW Somerset.
The weather experienced by Watford during the winter of 1978–79.
A study of social division in the Rickmansworth residential area.
A physical comparison of two valleys within the Pennines.
To examine the relationship between accessibility and functional hierarchy in the South Devon area, taking the South Hams region in particular.
The Burgess model of urban structure tested in NW London.

Part A

Collecting the information

When we hear geographical statements such as 'the average precipitation in London is sixty centimetres per annum', or 'the population of Newcastle has fallen by five per cent in the last decade', we tend to accept these as given facts. We forget that in order to obtain this information large numbers of people have been employed in collecting the data. Its reliability is therefore subject to human failings and methods which are not always perfect.

Geography depends on high quality information from which conclusions can be drawn. This section looks at some of the ways in which you can collect information and be certain that it is reliable.

Always bear in mind that there are two sorts of data available for most project work: 'primary data' which you collect yourself direct from the field (questionnaires, pedestrian counts or temperature readings, for example) and 'secondary data' which is available from someone else (such as the National Census or Water Authority data).

2 Sampling

2.1 Why do we usually sample?

If we wanted to examine the shopping habits of people in a middle-sized town it would not normally be possible to interview all the residents: we would have to 'sample' a carefully selected cross-section.

It is not always necessary to sample. If the total number of items under consideration, known in statistics as the 'population', is small then we might be able to measure all of them. Where the total population is less than 100 it would normally be wise to measure all of them: a sample of such a small number is likely to be unreliable.

But often sampling *is* desirable, for several reasons:

1 It is quicker.
2 It is cheaper.
3 It may be impossible to measure everything. For example, if we did a questionnaire survey of people's shopping preferences in a town, many would refuse to answer the questionnaire. We can only sample those who are willing to answer – and this, of course, makes the sample biased towards cooperative passers-by.
4 Where the population is changing quite rapidly we may wish to measure it at one moment in time only. Here we would be forced to sample because we could not get round in time to measure the whole population. For instance, because weather information is needed every day, meteorological readings are taken simultaneously at only a few 'sample sites' in the British Isles and sent immediately to the Meteorological Office. It would not be possible to record the weather at *every* spot in Britain simultaneously. To take another example, if we were recording the flow of traffic down streets in a local town it would be no good recording one street at 10 a.m. and another at noon because the two streets cannot then be fairly compared: traffic flows at noon may be quite different from those at 10 a.m. There would be a danger that any difference between the two streets was simply due to the different times at which flows were recorded and not to any difference in character between the two streets. For a fair comparison the traffic flows should be measured *simultaneously* in the two streets.

2.2 The sampling frame

The total number of items (the population) from which the sample is taken is called the 'sampling frame'. It is important to be completely clear about the nature of the sampling frame because this has a bearing on the reliability of the results of a sample taken from it.

Examples of possible sampling frames are:

1 A telephone directory, from which addresses can be sampled. The sampling frame is people who possess telephones and are not ex-directory. The sample would be biased against less wealthy people who do not have telephones.
2 House occupants who are interviewed during the day. This is biased against people who refuse to answer your questions and against working men and women who will not be at home during the day.

The precise characteristics of the sampling frame must be known if fair conclusions are to be drawn from the sample. Above all, one must avoid *bias*.

2.3 Sampling methods

Spatial sampling methods

When analysing the distribution of features (such as land uses, settlements and shops) three types of sampling method can be used (fig. 2.1):

1 **Point sampling** Individual points are chosen and the feature is sampled at those points. If sampling land uses, for example, one might select ten grid square intersections on an Ordnance Survey map and find out the land use at each of the ten points.

2 **Line sampling** Lines are drawn across the map and land uses (for example) along the line are noted.

3 **Area or quadrat sampling** Squares are chosen on the map or ground and the occurrence of the feature you are interested in is noted within those squares. This method is often used for sampling vegetation types, but it has been mathematically proved to be the least reliable of the three methods.

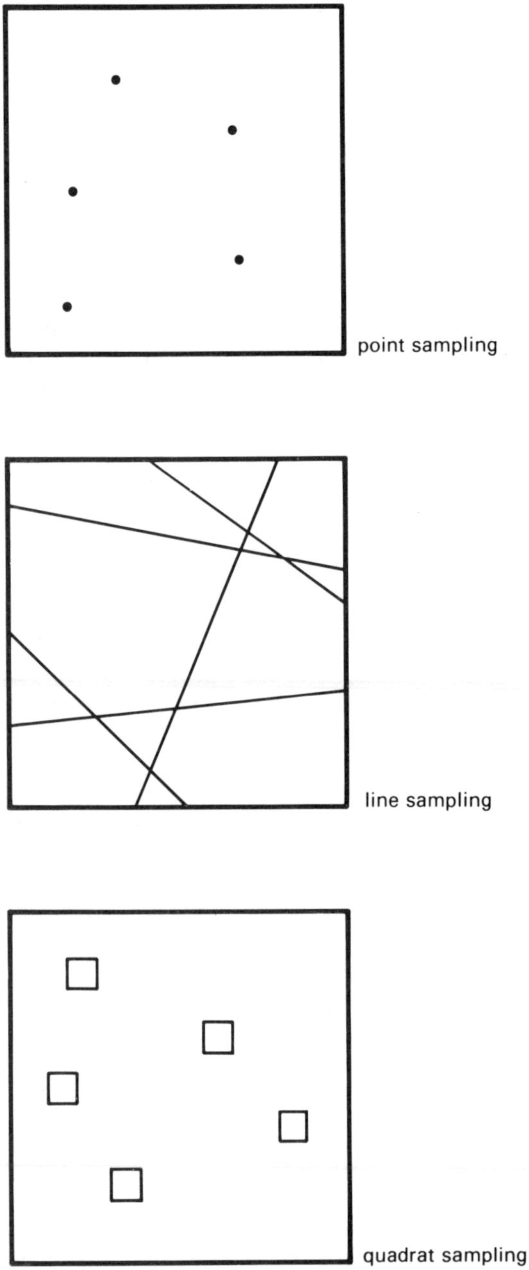

Fig 2.1 The three spatial sampling methods

General sampling methods

How can we decide where to put our points, lines or quadrats? It is no good just closing our eyes and stabbing at the map with a pencil because such a method may well result in bias. There are three commonly used methods. To illustrate these we will use in each case the example of an area composed of two rock types (A and B) on which we wish to sample land use (fig. 2.2). We will use the point sampling method but it would have been possible to use the line or quadrat methods instead.

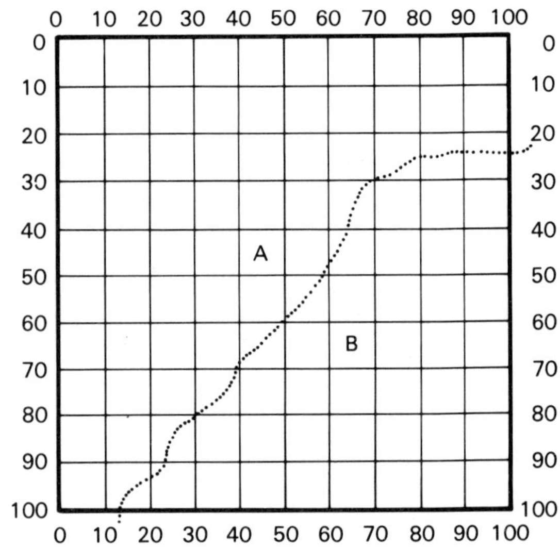

Fig 2.2 An area of land with two rock types, A and B. The map is covered by a grid.

1 Simple random sampling:

a Decide how many sample points you want.

b Take some random number tables (page 118). These are produced so that the numbers are totally random in their ordering. Read carefully

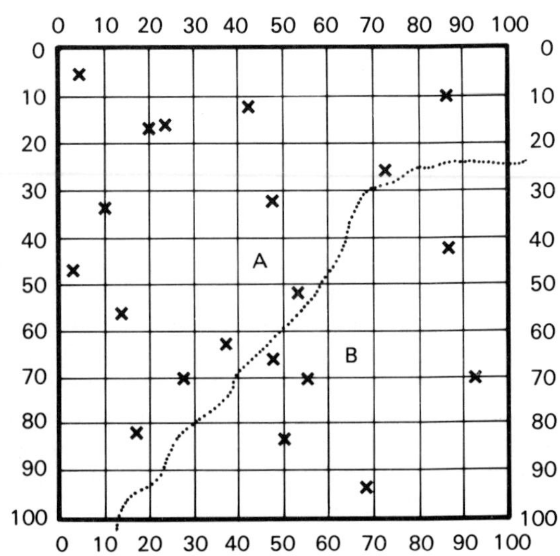

Fig 2.3 Random sampling, points method. 20 points are plotted on the map at random

the instructions on how to use them (above the tables).

c Cover the area of the map you are interested in with either ten, one hundred or one thousand grid squares and number them. If this is not done random numbers cannot be used correctly.

d Use the random number tables to read off grid references. Plot these on the map (fig. 2.3).

You will see that starting off in the top left-hand corner of the table on page 120 the first pair of numbers we come to is 20 17. Thus we take 20 as the easting grid reference and 17 as the northing grid reference. These are used to plot a point on figure 2.3. This procedure has been followed for twenty points.

e Find out the land use at each of the points chosen, either from a land use map or by visiting places on the ground. In practice this latter method may be difficult for it requires both accurate map reading and easy access to the place concerned.

2 Stratified random sampling In this method the population is divided into subsets and separate random samples are drawn from each subset. Because our example has two different rock types it is a case where stratified random sampling would be useful. We wish to sample land uses and rock type probably influences land use. Using simple random sampling we might be unlucky enough to get all our sample points on just one of the rock types, omitting any consideration of the possible influence of the other rock and producing a biased result. It is thus desirable to get a complete cover of the whole area with a proportionate number of sample points occurring on each rock type. The method is as follows:

a Decide how many subsets (or strata) you need. In this case there are two rock types so we need two strata; if there had been three rock types we should have needed three strata or subsets.

b Decide how many sample points you need. In our example we have decided to sample twenty points.

c Divide this number amongst the strata in proportion to their importance. In our example rock type A takes up 60% of the area and will thus receive 60% of the sample points, i.e. twelve points (fig. 2.4). The other eight points will be allocated to rock type B which takes up 40% of the area.

d Proceed as for simple random sampling (stages b–e). Do not allow any more than the alloted number of points to fall on each stratum. Ignore any excess points which may arise.

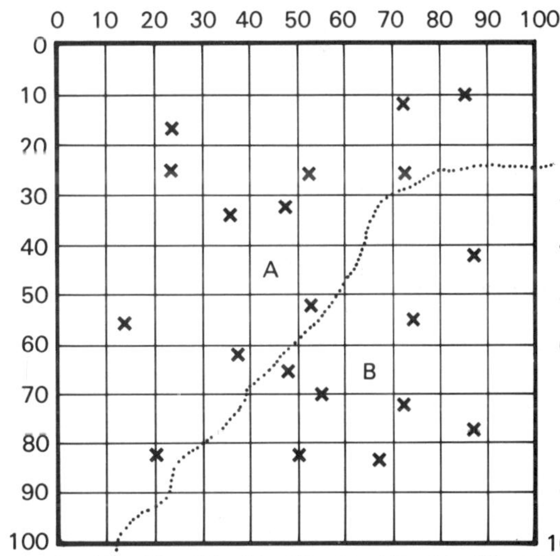

Fig 2.4 Stratified random sampling. 20 points are plotted on the map, 40% (8 points) in area B which is 40% of the total area

3 Systematic aligned sampling In this method the sample points are located in a regular or systematic fashion across the map or area. The method is simple:

a Decide how many sample points you need.

b Place these points regularly over the map (fig. 2.5).

c Find out the land use at each point, as for stage (e) in simple random sampling.

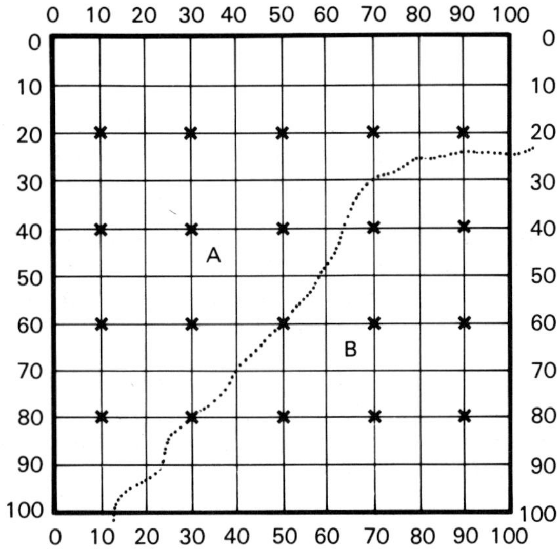

Fig 2.5 Systematic aligned sampling. 20 points are plotted on the map in a regular fashion

The advantages of this method are that it ensures a complete cover of the area (like stratified random sampling) and is simple to do. The danger is that the systematic sample might inadvertently pick up some systematic occurrence of a phenomenon on the map or area. For example, systematic sampling of this kind would lead to problems if done on North American cities, which tend to be divided into regular blocks, for it might easily pick up some regularity on the ground: for example, all the points *could* land on street corners, which tend to possess banks or newsagents.

These three general sampling methods can be applied to any type of sampling. If sampling names from a telephone directory, for example, we could pick numbers from the random number tables and count down that number of names from the top of the page (random sampling). Alternatively, we could pick every twentieth name (systematic sampling).

If doing line sampling we could pick grid references from the random number tables, one for the x-axis (horizontal axis) of the grid and one for the y-axis (vertical axis) and join these to give a line (random sampling). We might do the same again, ensuring that this time one line appeared in each half of the map (stratified sampling). Or we might place the lines regularly down the map (systematic sampling).

Sample size Usually the size of sample you take will be dictated by the amount of time and resources available. It is important to realise that the larger the sample the more likely it is to give you a true picture of the population you are sampling. So the general rule is 'the more the better'.

Thinking it through Try to think carefully about your sampling method before you begin – the details matter. For example, if you wanted to sample pebbles on a beach you might decide to lay a tape-measure across it and pick up a pebble every ten metres. But which pebble should you choose? One *touching* the 10 m mark on the measure? What about pebbles just beneath the surface?

If you were going to interview people in the High Street, which days of the week should you do it on? At what times of day? Where exactly should you stand? And how do you choose which people to interview?

Your answers to these questions will determine the sort of results you get. This is why you have to decide beforehand exactly what you are going to do and *why*.

To summarise, here is a check-list of the decisions which must be made when sampling:

1 Do I need to sample? If so . . .
2 What is my sampling frame? Is it biased?
3 Do I need to sample spatially (on a map or on the ground)? If so, shall I use point, line or quadrat sampling?
4 Shall I use simple random, stratified or systematic sampling methods?
5 How many samples shall I collect?

2.4 Exercises

1 Figure 2.6 shows nine different spatial sampling methods. For each map state whether the method used is point, line or quadrat and random, stratified or systematic.

2 a Obtain a Land Use Survey map. Randomly select twenty grid intersections. Note the land use at each. This will allow you to estimate the percentage of the map area devoted to different forms of land use.

b Repeat the above procedure, this time using five systematically aligned lines. Note the proportion of the various land uses along each and estimate the percentage of the map area devoted to the different land uses. How does your result differ from that of (a) above?

3 You wish to do a survey of the shopping habits of people living in a small town. Describe as exactly as you can the various methods you could use to select a sample of people for this survey.

4 You wish to do a survey of pebble sizes on a beach. Describe as exactly as you can the various methods you could use to select an appropriate sample of pebbles for this survey.

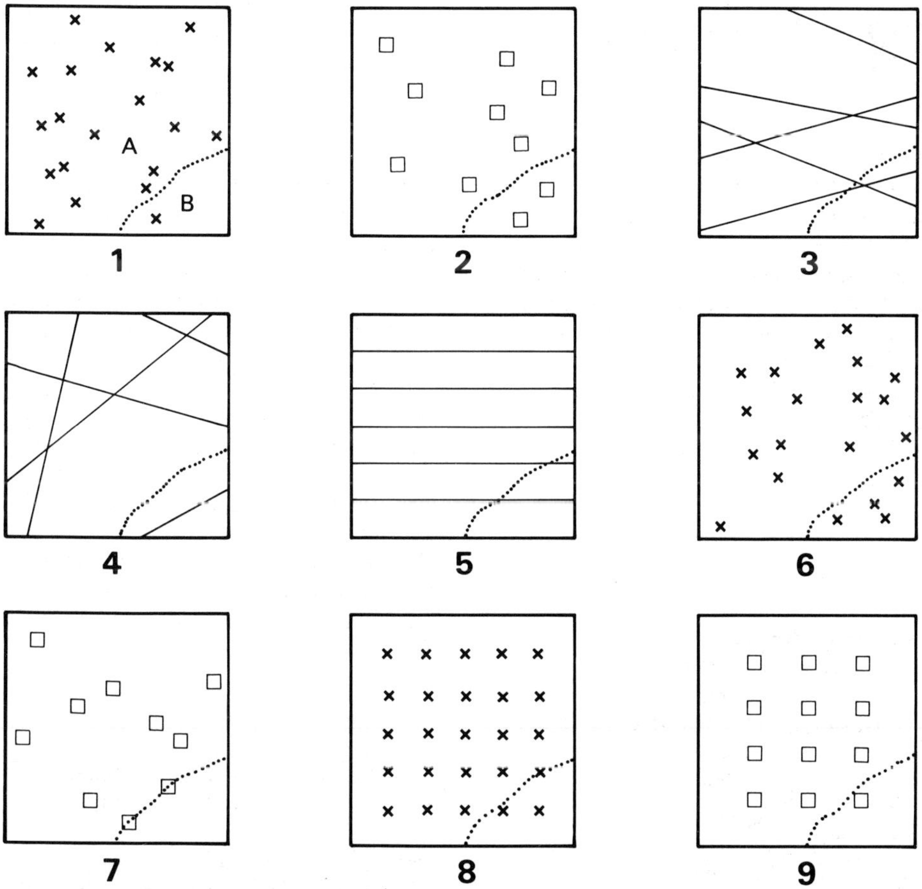

Fig 2.6 Nine sampling methods which are used to determine land use. The area is divided into two rock types (A and B in 1) and rock type influences land use.

3 Physical geography

This chapter is broken down into a number of sections each dealing with a separate aspect of physical geography. Techniques for the collection of data are described, followed by suggestions and advice on projects.

3.1 Geology, relief and slopes

Whatever aspect of physical geography you may be investigating it is often valuable to know something about the general relief, the forms of the slopes and the underlying geology in your study area. This section describes some of the ways in which topography can be measured and some secondary data sources from which details of geological structure and lithology may be obtained.

Geology

A knowledge of rock types is necessary in order to understand landforms and landscapes. In many cases you will have no difficulty in identifying rock samples, but in areas of complex geology, maps of solid and drift geology may be useful. Geological maps are obtainable from good bookshops or from the Geological Museum, Exhibition Road, London SW7 2DE. Maps for parts of Great Britain are available at the approximate scales of 1:625 000, 1:250 000 and 1:10 000. More complete coverage exists at the 1:50 000 scale, although even here there are gaps.

Weathering

The two main types of weathering are 'physical' and 'chemical' weathering. Physical weathering is caused by stresses created in rocks by heating and cooling, the freezing and thawing of water, and wetting and drying. Chemical weathering occurs when minerals in rocks are attacked by weak acids in rain water.

Distinctive landforms associated with different rock types develop partly as a result of the types and intensities of weathering processes to which they have been subjected. Rates of weathering are controlled by climatic factors (latitudinal and maritime influences), microclimatic conditions (altitude and aspect) and the susceptibility (in terms of both chemical composition and physical properties) of different rocks to weathering processes. Because weathering takes place at very slow rates accurate, quantitative measurements cannot be undertaken without expensive equipment. But it is possible to make qualitative, subjective assessments of the degree to which rocks have been weathered. A visual technique for determining the extent of weathering can be used, for example, on gravestones. Since gravestones are dated the approximate time taken to arrive at a particular state of decay may be calculated. Some categories of degree of weathering are given in figure 3.1.

Class	Description
1	Unweathered
2	Slightly weathered – faint rounding of corners of letters
3	Moderately weathered – rough surface, letters legible
4	Badly weathered – letters difficult to read
5	Very badly weathered – letters almost indistinguishable
6	Extremely weathered – no letters left, scaling

Fig 3.1 A visual classification of the extent of weathering on gravestones

An index of the amount of *chemical weathering* taking place in an upland area may be determined in the following way:

1 Measure the concentration of solution load in a stream with a conductivity meter (described on page 27).
2 Measure stream discharge (described on page 20).
3 Take these readings together every day for a few days and calculate mean conductivity and mean discharge (see page 91 for calculation of mean values). Do not take readings in a period following heavy rain, as discharge will be unusually high and may give misleading results.
4 Multiply the two values together to determine the total number of conductivity units (L), a measure of total dissolved load.
5 Calculate the area which the stream drains by referring to an Ordnance Survey map and adding the number of grid squares the basin covers. Each grid square of an Ordnance Survey map is 1 km^2.
6 Divide the value (L) by the total area to obtain the

Region 1	
Conductivity of stream (C)	= 45 μ mhos
Discharge of stream (Q)	= 30 litres/second
Total dissolved load (L)	= 1350 (i.e. $C \times Q$)
Drainage basin area (a)	= 4 km²
Index of chemical weathering (L/a)	= 3.38
Region 2	
Conductivity of stream (C)	= 15 μ mhos
Discharge of stream (Q)	= 60 litres/second
Total dissolved load (L)	= 900
Drainage basin area (a)	= 15 km²
Index of chemical weathering (L/a)	= 0.6

Fig 3.2 Calculation of an index of chemical weathering

index of chemical weathering. A worked example is shown in figure 3.2. The index is used purely for comparative purposes. It does not tell us, for instance, the rate of surface lowering, but will allow us to compare drainage basins and identify the one with the highest rate of chemical weathering.

Slope surveying

The end product of a slope survey is either a slope profile (fig. 3.3) or a map. A profile is made up of many separate slope units each of which must be surveyed. Measure the length of each slope unit with a tape and find the slope angle with a clinometer or Abney level. These instruments can be bought although details for making a clinometer are given on page 22. The method for using such instruments is as follows:

1 Two people each hold one pole vertically with the bottom of the pole resting on the ground surface, and with one pole at each end of the slope unit to be measured.

2 The instrument is then sighted from a chosen height on one pole to the same level on the other pole and the angle of slope is read off.

Slope measurements

Slope reading	Length of slope measured (m)	Angle of slope (°)
1	10	13
2	15	2
3	12	29
4	11	6

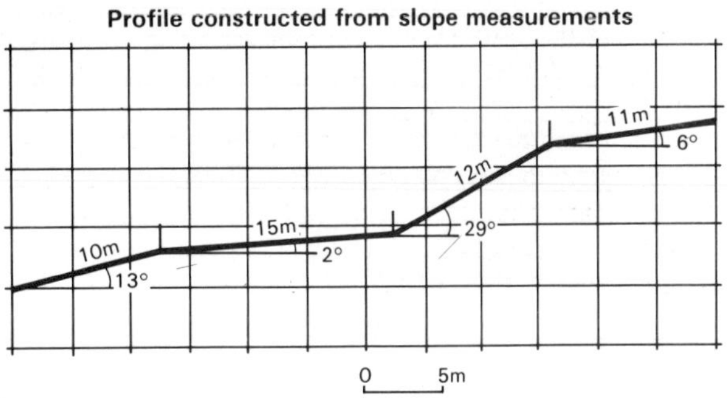

Fig 3.3 Slope profile of valley side

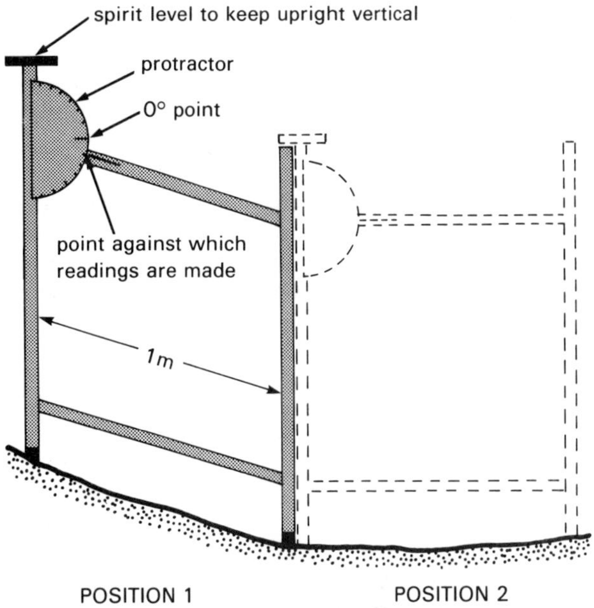

Fig 3.4 A slope pantometer. The instrument is stepped downhill and at each position the angle made by the upper cross-piece with the protractor is noted.

3 The distance between the two poles is measured. The shorter the distance between the poles the more accurate the final result will be. Measurements are usually taken at each break in slope.

4 The resulting measurements are then plotted on graph paper to reconstruct the slope profile (fig. 3.3), bearing in mind that the length measurements are along the slope, *not* horizontal.

Alternatively you can make your own surveying instrument, a 'slope pantometer' (fig. 3.4). Two uprights (1 m apart) are bolted loosely together with two cross-pieces. A large protractor and a spirit level are fixed to one upright. The pantometer is placed on the slope and, using the spirit level to keep the uprights vertical, the angle made by the protractor with the upper cross-piece is noted. The pantometer is stepped downhill perpendicular to the slope contours and the angle noted at every step (i.e. every metre). This technique is rather slower than the other one but has the advantage that only one person is required.

The value of slope surveying is that it enables you to make a detailed study of the profile of the slope. Slope profiles reflect both the underlying geology and past and present processes acting on them.

Field sketching

A field sketch is exactly what it says, a sketch of a landform or landscape made in the field. An example is shown in figure 3.5. The value of a field sketch is that it forces the observer to look in detail at the features before him and is of course cheaper than a photograph. The sketch should be kept simple, bringing out the important geographical elements of the landscape, and should be labelled.

Fig 3.5 (a) Stair Hole and Lulworth Cove, Dorset

The most reliable method is to draw two lines across the page, dividing the sheet into three. This prevents the common error of too much vertical exaggeration and leaves room for labelling. Using a pencil draw in the skyline, then a line for the foreground. Fill in the middle, including major but excluding minor features.

We suggest that you practise your field sketching technique looking at slides in the classroom before going out into the field.

Morphological mapping

Morphological maps are maps of slopes and the topography, or shape, of the land. They are drawn by observing the terrain and measuring slopes in the ways described on pages 11–12.

An example of a morphological map is shown in figure 3.6. Such maps are useful when studying any geomorphological feature where slope form and slope processes are important, such as slumps, landslides,

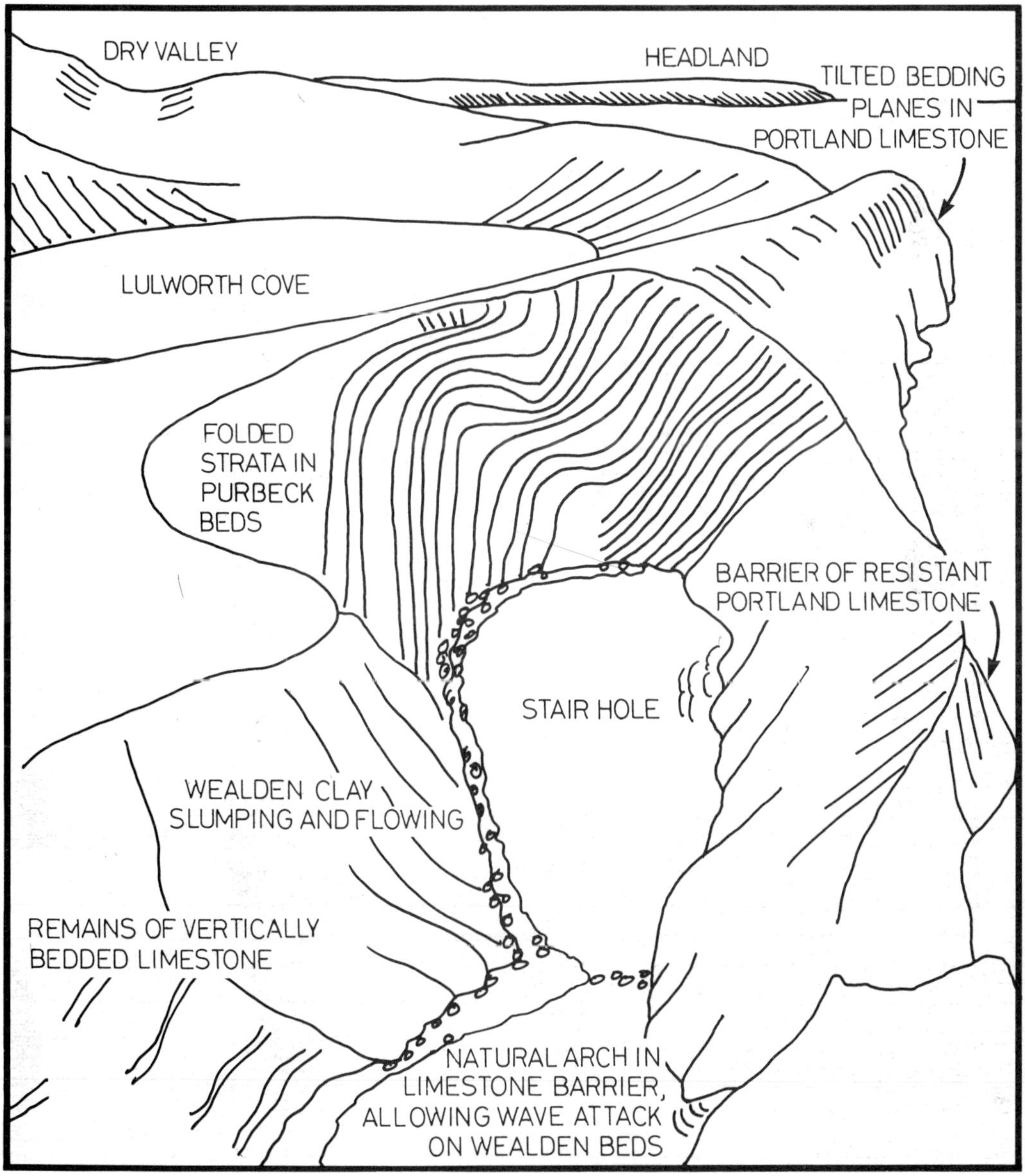

Fig 3.5 (b) Field sketch of Stair Hole and Lulworth Cove, Dorset

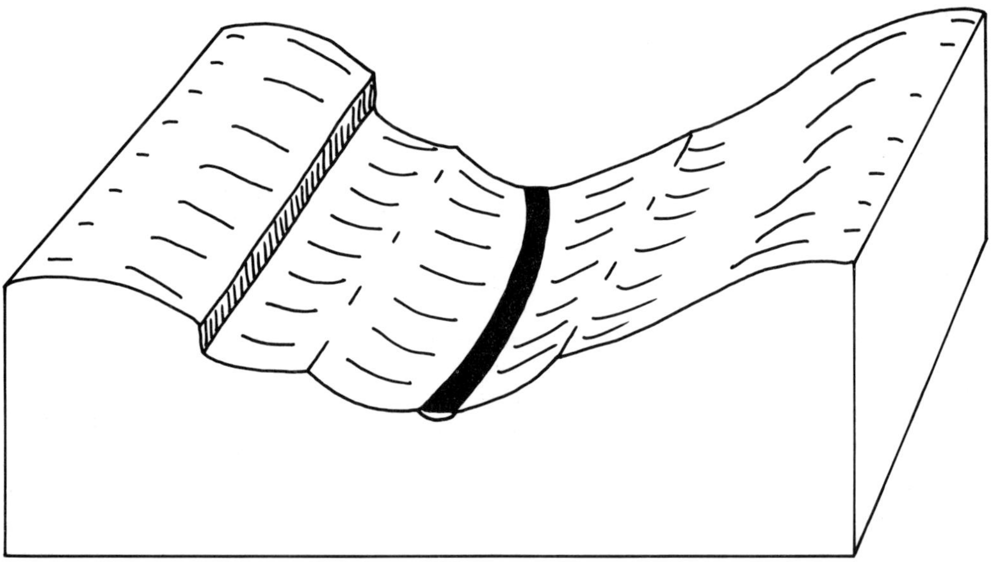

Fig 3.6 (a) Three-dimensional view of landscape

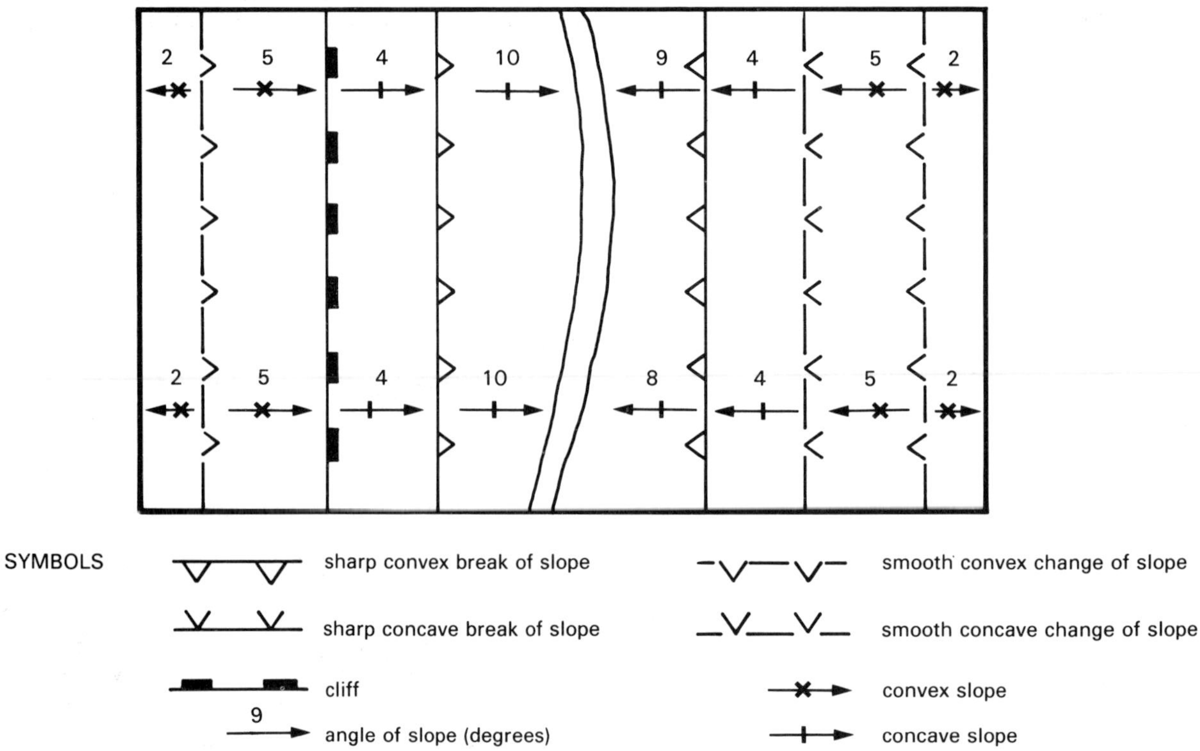

Fig 3.6 (b) Morphological map of landscape

river terraces and beaches. Morphological maps are of great value to those planning roads and building new settlements where slope angles are critical.

Project suggestions

Many worthwhile projects may be based on the techniques described above:

1 Investigate the amount of weathering that has taken place on gravestones of different geology. In a large graveyard with many tombstones you could relate degree of weathering (using figure 3.1) to rock type, age of gravestone and aspect.

2 Calculate the index of chemical weathering for small upland streams draining areas of different geology. Compare the indices and relate the values to the susceptibility of the different rock types to chemical weathering processes.

3 Another line of inquiry in an area of contrasting geology could be to measure the steepness of slopes on different rock types. Measure slope angles on Carboniferous Limestone, Millstone Grit and shales, for instance. Plot the data as a series of bar graphs and conduct a test of significance (page 98) to see whether differences in slope angle could be accounted for by rock type or are merely a chance occurrence. Possibly different processes operate on different rock types; certainly stream flow is more prevalent on clay than on chalk owing to the greater permeability of chalk. These different processes may be reflected in different angles of slope.

4 Relating geology, relief and land use in a local area can also be interesting. To do this particular project you would need to survey the slopes and produce a morphological map. Map local land use and compare and relate both of these with an official geological map or one you have produced yourself.

3.2 Hydrology

Hydrology is the study of water and its continual movement through the hydrological cycle (fig. 3.7). The hydrological cycle can be considered at different scales: global, regional or at the level of the individual drainage basin. You are most likely to do project work at this last scale.

Both the amount of water entering a basin as precipitation ('input') and leaving as streamflow or by evaporation and transpiration ('output') can be measured. By comparing inputs and outputs hydrologists can calculate the net loss or gain of water to a basin over a given time. Any precipitation that does not leave within the period under observation must have been *stored* within the basin either in the soil or

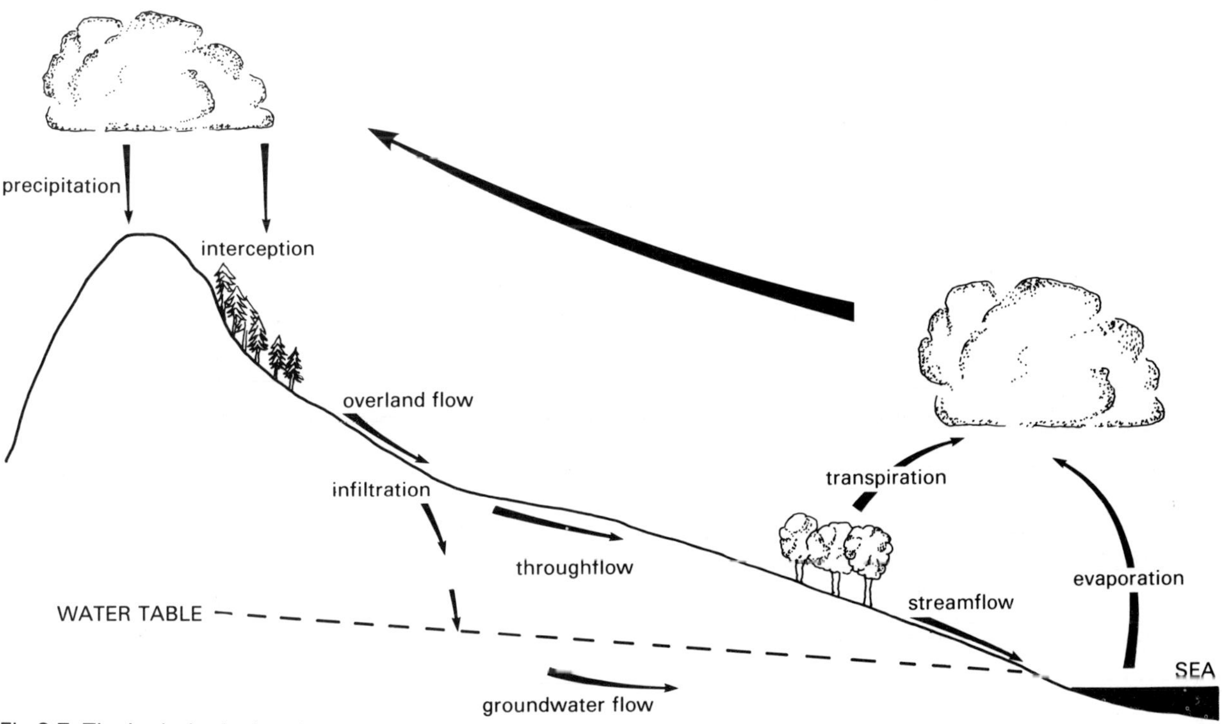

Fig 3.7 The hydrological cycle

in the rock. Similarly any water leaving that cannot be accounted for by recent rainfall must have been drawn from stores of water in the basin. Even during a very dry summer many rivers continue to flow, fed by water draining from aquifers (permeable rock strata). (Note: the terms *river* and *stream* are used synonymously in this book.)

The techniques described in this section include most of the methods by which you can investigate the hydrology of a small area.

Inputs

Precipitation

The input of water to a drainage basin in the form of rain or snow can be measured with a rain gauge (page 45). However this only tells us the amount of water falling at *one* point. A rain gauge is effectively a point sample (page 5) and we have to assume that the same amount of rain falls over the surrounding area. Over a large area several rain gauges may be used and an average figure for the region calculated using Thiessen polygons (fig. 3.8).

Not all the rain water is able to make its way to the stream, particularly if the ground is densely vegetated or wooded. If rain lands on trees it is said to have been 'intercepted'; some may be stored on leaves and will eventually be evaporated or will reach the ground via the trunk or by dripping straight from the leaves. The amount of rain intercepted in a wooded area may be calculated by placing several rain gauges beneath trees and one or two more in a nearby open space. The difference between the average amounts collected will be the amount intercepted. Obviously this figure will vary according to the type of tree and time of year.

Movement of water within the drainage basin

Infiltration rate

On reaching the ground, water enters the soil. The speed at which this occurs is known as the 'infiltration rate'. Infiltration rates depend on the degree of compaction of the ground surface, soil particle size and the existing soil moisture content. Plant roots help keep the soil structure loose so the presence of vegetation promotes infiltration. Land use also affects soil compaction. The soil of a heavily used area of land will be more compact than that of a lightly used area. The more compact the ground, the lower the infiltration rate. A track across a field would accordingly have a considerably lower infiltration rate than a remote corner of the field. If recent weather conditions have been wet many pore spaces in the soil will be waterlogged and this will reduce the capacity of the ground for absorbing more water, thus favouring increased surface runoff.

The standard technique for calculating rainfall over a large area, given the data from a network of rain gauges, is to construct and use Thiessen polygons:

1 Plot location of gauges on a large-scale map.
2 Join gauges with straight construction lines.
3 Bisect construction lines with perpendiculars to produce polygons.
4 With the aid of transparent graph paper, calculate the area of each polygon.
5 Mean rainfall for whole basin

$$= \text{Depth of rain in gauge} \times \frac{\text{Area of polygon}}{\text{Total catchment area}}$$

$$= \left(10 \times \frac{3}{7.5}\right) + \left(15 \times \frac{2}{7.5}\right) + \left(12 \times \frac{2.5}{7.5}\right)$$

$$= 12 \text{ mm}$$

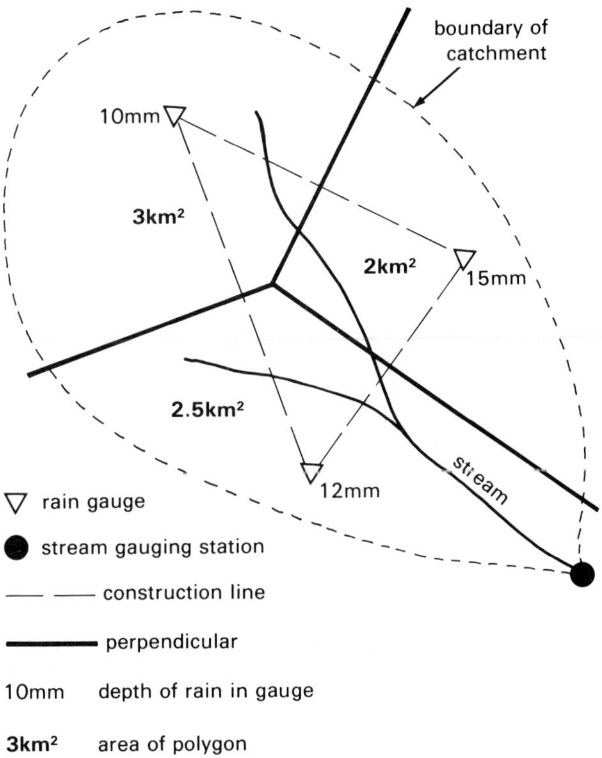

Fig 3.8 Calculation of rainfall using Thiessen polygons

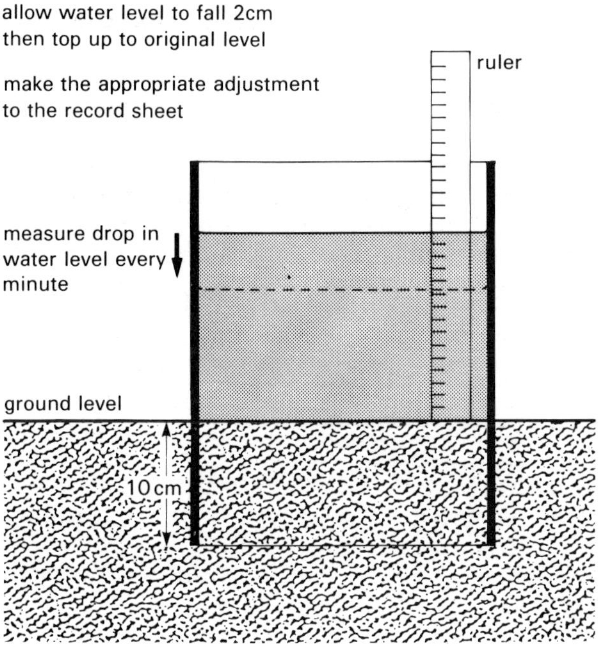

Fig 3.9 Measuring infiltration rate

Fig 3.10 Infiltration ring

The method for measuring infiltration rates is as follows:

1 Obtain a large cylinder (at least 30 cm tall). A large tin can, such as a catering size can which you could get from your school kitchens, is ideal. Remove the top and bottom.
2 Make sure that you have a plentiful supply of water, a watch with a second hand and a clipboard with a pencil and paper. At least two people will be needed to take the readings.
3 Hammer the ring into the ground to a depth of about 10 cm on your chosen site (fig. 3.9).
4 Place a ruler vertically inside the ring to record the fall in water level.
5 Pour water into the cylinder to a depth of 15 cm above the ground (fig. 3.10). At first the water will be absorbed quickly, so record the drop in water level every minute. If the rate slows you may take readings less frequently.
6 As the water level falls then so too will the water pressure. This alone will produce a lower infiltration rate, thus giving misleading readings. To overcome this problem allow the water to fall 2 cm then top the ring up with water, recording the exact depth added and making the appropriate adjustment to your record sheet.
7 Plot the results on a graph (fig. 3.11).

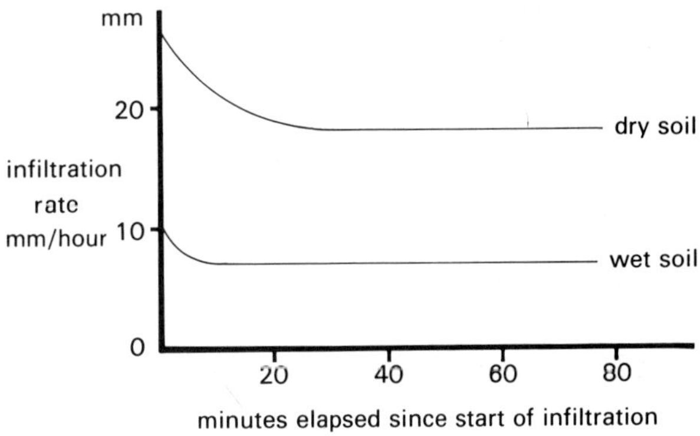

Fig 3.11 Infiltration curves

Overland flow

This occurs when the ground is wet and the rain so intense that the infiltration rate is exceeded. The ground can no longer absorb water so the excess flows over the surface unchannelled. This is hard to measure and quite rare in the UK except during heavy storms. The only places at which you are likely to encounter this type of flow are at the foot of a slope or near a river channel, and then only under exceptionally wet conditions. (It also occurs on impermeable rock surfaces, though it cannot be measured by the following technique.) The technique for measuring overland flow is as follows:

1 Dig a trench 5–10 m long across the slope, about 30 cm deep and 30 cm wide.
2 Secure a length of guttering flush with the upslope side of the trench to ensure that all overland flow will be collected. The gutter needs to be at a slight angle to allow the water to drain into a measuring jug at one end (fig. 3.12).
3 Cover the trench with a polythene sheet to keep out the rain.
4 When overland flow occurs, measure the amount collected in the jug at hourly intervals. At the same time measure rainfall so that you can get an idea of the conditions that lead to the occurrence of overland flow.

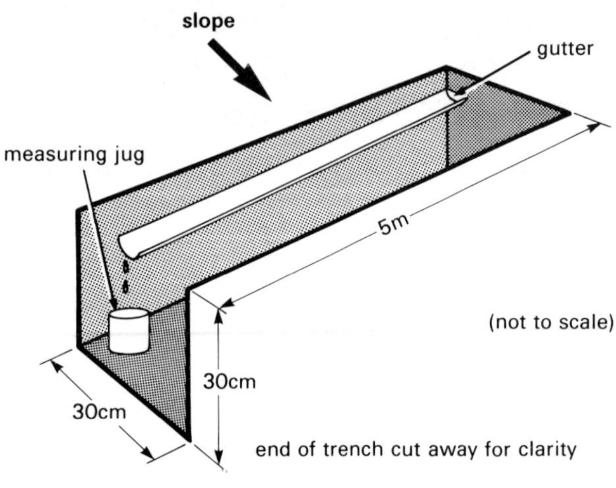

Fig 3.12 Trough for measuring overland flow

Throughflow

Water flowing laterally through the soil beneath the surface is called throughflow. It is measured in much the same way as overland flow. The main difference is that the trench needs to be deeper and you need to install a second gutter 50 cm beneath the one at the top of the trench measuring overland flow (fig. 3.13). The first gutter must be retained to prevent overland flow from running into the throughflow gutter.

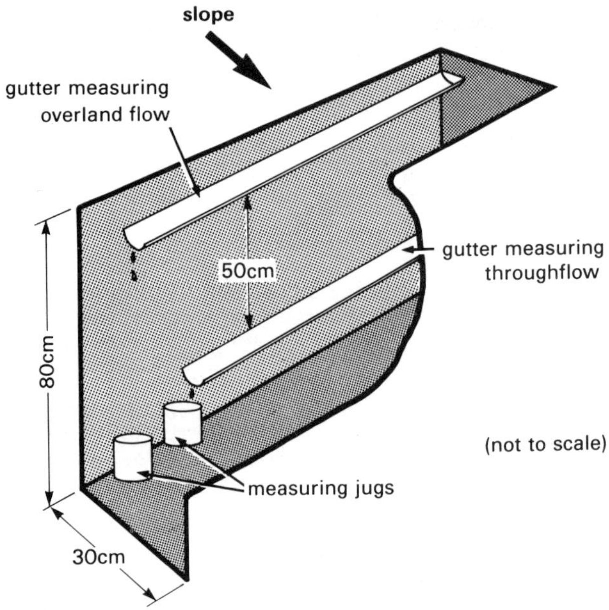

Fig 3.13 Trough for measuring throughflow

Note the 'lag time' between a rainstorm and the emergence of throughflow. This can be related to such influences as vegetation cover and infiltration rates by comparing lag times from different areas.

Groundwater

Water in the soil may eventually drain down vertically into the rock beneath. If the rock is sufficiently permeable the water will remain stored for months or even years and is termed 'groundwater'. Accurate measurements of the volume of groundwater are very difficult to obtain although you can record changes in the height of the water table (the upward limit of the groundwater) by monitoring the level of water in wells. This allows us to say that the store of water in the ground has increased or decreased over a given period, but no more. We cannot measure the change in, say, thousands of litres.

Outputs

Evaporation

Water is lost from drainage basins in many ways. Any basin with a lake or other expanse of open water will lose significant amounts of water by evaporation.

You can measure the approximate rate at which water is lost from a lake using the following technique:

1 You need an evaporation pan – a large open container from which you can record the rate of water loss. A large plastic washing-up bowl or biscuit tin is suitable, but it must not have slanting sides. Place the pan on the ground and fill it with water (fig. 3.14).
2 Measure the depth of water in the pan at frequent intervals (every few hours on a hot day). Keep it topped-up to the original level. If it rains you must subtract the recorded rainfall figure, so try and record the depth of water at the beginning of the storm and again at the end and subtract the difference. The result can be expressed in millimetres per day.
3 If the pan being used to assess water loss is less than 3 m in diameter, the evaporation loss will be greater than from a large body of water as the temperature of the pan sides affects readings. Therefore all results must be reduced by 25% to compensate.

Fig 3.14 Evaporation pan

Transpiration

Water transpired from plant leaves to the atmosphere (especially in the summer in a densely vegetated basin) represents a large proportion of the total output of a drainage basin. A single mature oak tree can lose thousands of litres of water on a hot day. Obviously different types and sizes of plants transpire at different rates. Trees, with deep roots which can reach down for moisture, will continue to transpire when grass, with short roots, has begun wilting due to lack of available water. Thus within a drainage basin transpiration rates vary spatially with changes in vegetation type. Rates of water loss from vegetation also vary over time with changes in the weather.

The technique described below, although somewhat crude, will give you some idea of the rate at which water is lost:

1 Select a leafy branch. Cover it with a large, clear polythene bag, tied at the base to make it watertight.
2 After a given interval (e.g. one day) return with a graduated cylinder and measure the amount of water collected inside the bag (fig. 3.15).
3 The total volume recorded can then be related to weather conditions, in particular temperature and sunlight, although of course the bag will protect the plant from drying wind, thus introducing inaccuracy.

Transpiration rates can also be measured using filter paper which has been soaked in cobalt chloride solution. Dry the paper out and clip a small piece

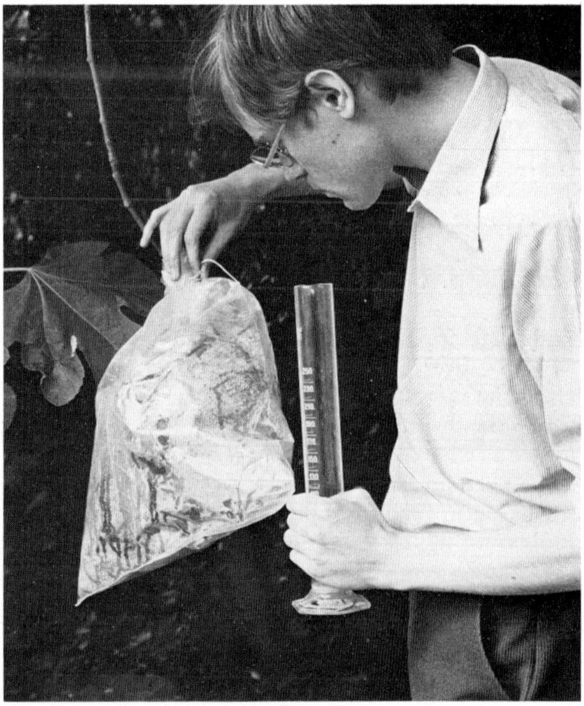

Fig 3.15 Measuring the water transpired by a tree branch into a polythene bag

onto the leaf of the plant between two glass slides. At this stage the paper is blue but on absorbing water transpired by the plant it will turn pink. Record the time taken for this colour change to occur: this is a measure of the transpiration rate which enables you to compare different types of vegetation.

Stream discharge

Discharge is one of the more easily measured components of the hydrological cycle. It is defined as the volume of water passing a given point on a river bank in a given time. The volume is expressed either in cubic metres per second (cumecs) or in litres per second. The most commonly used technique for measuring discharge is the velocity–area method:

To calculate stream velocity:

1 Select a straight reach of river 10–30 m long, preferably with no pools, eddies or waterfalls. Measure the distance (*d*) with a tape.

2 Obtain some floats whose velocity you can record over the measured reach. The best are those which float largely beneath the surface and are thus unaffected by wind, for example dog biscuits or oranges.

3 Accurately time the float over the measured distance at least three times and calculate the average. If the channel is sufficiently wide it may be possible to calculate the velocity in the centre of the river and also towards the banks.

4 Multiply the average time by 0.85, because the water on the surface flows faster than that beneath and this conversion ensures an accurate velocity reading for the whole cross-section.

5 Now calculate the velocity:

$$\text{velocity (metres per second)} = \frac{\text{distance (metres)}}{\text{time (seconds)}}$$

See figure 3.16 for a worked example.

To calculate cross-sectional area:

1 Measure stream width with a measuring tape.

2 Hold the tape taut across the stream and measure water depth at regular intervals across the width (e.g. every 25 cm in a 2 m wide stream). See figure 3.17. Take into account the fact that the water will splash up around the rule.

3 Plot your figures on graph paper to reconstruct the channel cross-section and calculate the area (in square metres). See figure 3.18 for a worked example.

Stream discharge can now be calculated by multiplying stream velocity by cross-sectional area (fig. 3.18). Figure 3.19 is an example of a data sheet that you could copy and use for collecting your discharge data.

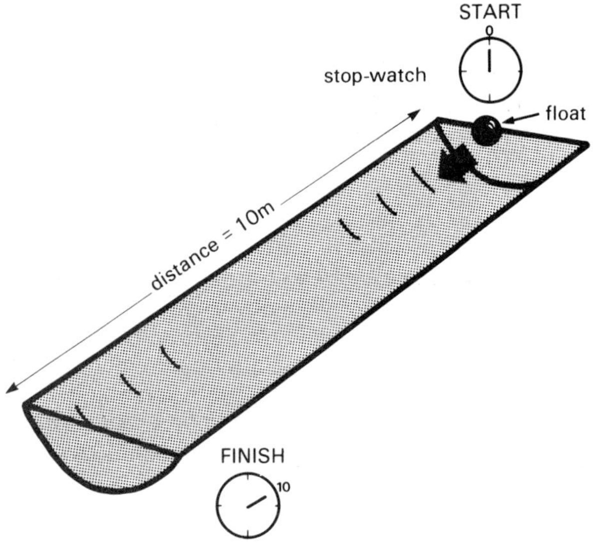

Average time of floats = 10 seconds
Multiply time by 0.85 = 8.5 seconds
distance (*d*) = 10 metres

$$\text{velocity} = \frac{\text{distance}}{\text{time}} = 0.85 \text{ m/sec}$$

Fig 3.16 Measurement of stream velocity

Fig 3.17 Measuring stream depth

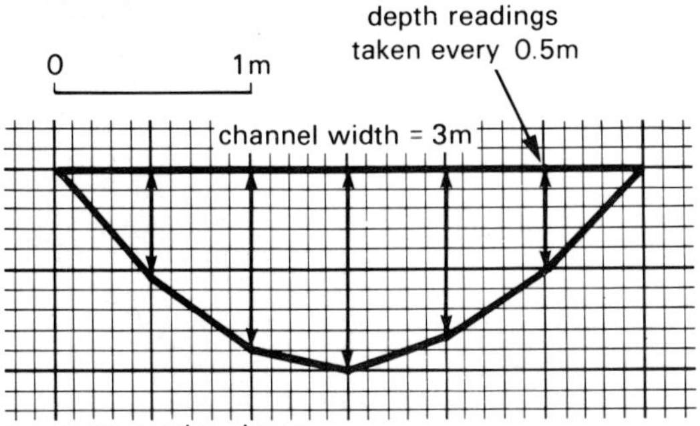

stream cross-section drawn
from measurements
taken on the stream

Reconstructed cross-section comprises 8 large graph paper squares.
Each large square is equivalent to 0.25 m^2

∴ stream cross-sectional area = 8 × 0.25
= 2 m^2

discharge = stream velocity × cross-sectional area
= 0.85 × 2
= 1.70 m^3/sec (cumecs)

Fig 3.18 Measurement of stream cross-sectional area and discharge

Project suggestions

There are many aspects of drainage basin hydrology that lend themselves to investigation:

1 How does discharge respond to a period of heavy rainfall? Try and relate discharge fluctuation not only to total input of rain but also to geological and vegetational influences.
2 How do rates of evaporation and transpiration fluctuate with daily weather changes?
3 It may be possible to investigate the hydrology of a small area, for example an orchard. Measure precipitation, evaporation, transpiration, infiltration rates and throughflow. Careful measurement of each of these components over a period of two weeks would allow you to calculate and compare the volume of inputs and outputs in your area.

If it is impossible to devote an uninterrupted period of time to your project you would be best advised not to tackle one of the above 'time based' projects but instead to undertake a 'spatially oriented' project. In other words rather than observe hydrological changes over time, see how they vary from place to place.

4 Measure rates of transpiration from different types of vegetation; compare deciduous and coniferous trees for example.
5 Compare infiltration rates under woodland and on grassland.

DISCHARGE DATA SHEET

DATE: ....

NUMBER/NAME OF REACH:...................................

1 MEASUREMENT OF VELOCITY

Length of reach studied .... metres
Time taken for first float .... seconds
Time taken for second float .... seconds
Time taken for third float .... seconds
Average time of floats .... seconds
Multiply by 0.85 .... seconds

Average velocity = $\frac{\text{distance}}{\text{time}}$ =............ $\frac{\text{metres}}{\text{seconds}}$ = ...m/sec

2 MEASUREMENT OF CROSS-SECTIONAL AREA

Channel width = ... metres
Channel depth at .... centimetre intervals = 1 .... 6 ....
2 .... 7 ....
3 .... 8 ....
4 .... 9 ....
5 .... 10 ....

Cross-sectional area = ... m^2

3 DISCHARGE

Discharge = velocity x cross-sectional area

= .... m^3/sec

Fig 3.19 Field sheet for discharge data

3.3 River channels

So far we have considered rivers as part of the hydrological cycle. In the previous section we described the measurement of river discharge – the principal loss of water from most drainage basins. This chapter describes how you can measure river channel size and shape and river load.

River channel form

A good impression of channel size and shape can be gained by taking the following measurements.

The *gradient* of a river is measured with a clinometer. This instrument indicates angles of tilt and may be obtained commercially or can be made (fig. 3.20). The measurement of stream gradient is illustrated in figures 3.21 and 3.22. By recording gradient at close intervals the long profile of a reach may be reconstructed as figure 3.22 shows.

Channel *width* is measured with a rule or tape-measure and channel *depth* with a rule. Distinguish between channel dimensions and those of the flowing water. In a dry summer streams may be many times smaller than their channels. If you are investigating the channel (as opposed to the stream), measure the

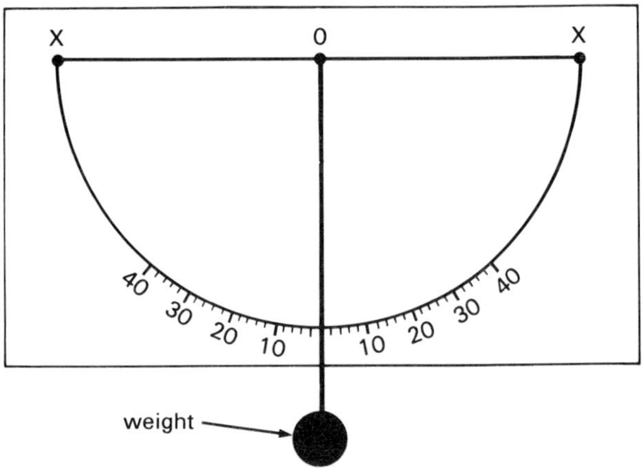

1 Trace a protractor onto a sheet of paper and mark on degrees as above.
2 Glue the paper to a sheet of card.
3 Thread some string through a hole at the point marked 'O'. Secure with a knot and fix a weight to the loose end.
4 Push two drawing pins through the points marked 'X' from behind. These serve as sights to help align the clinometer.

Fig 3.20 Construction of a clinometer

Fig 3.21 Measuring stream gradient

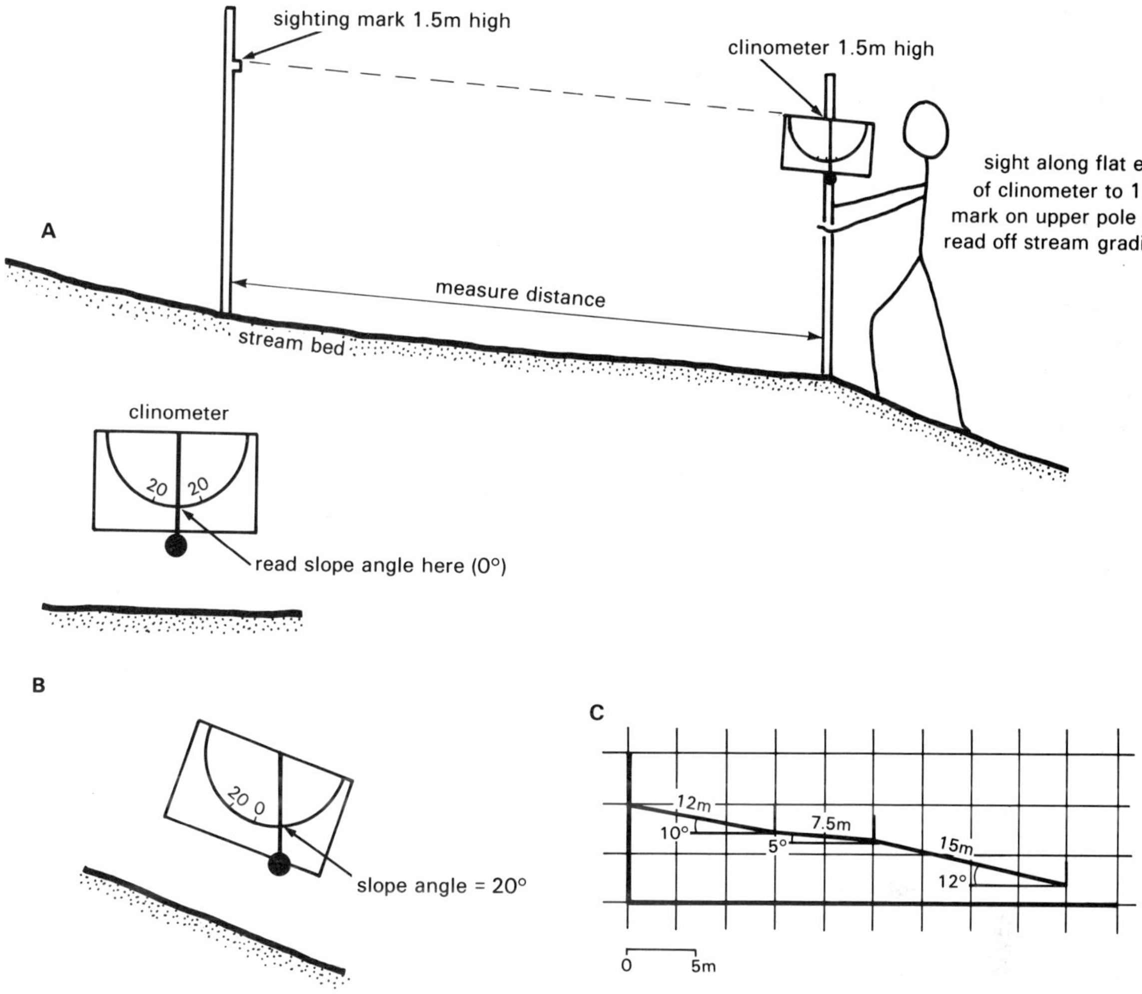

Fig 3.22 Measuring stream gradient with a clinometer

dimensions that would be occupied by water at bankfull discharge as figure 3.23 illustrates. Bankfull discharge is significant because when rivers are at this stage channel features such as meanders are formed and most erosion of bed and banks takes place.

Another aspect of channel form which helps explain observed fluvial processes is 'channel efficiency'.

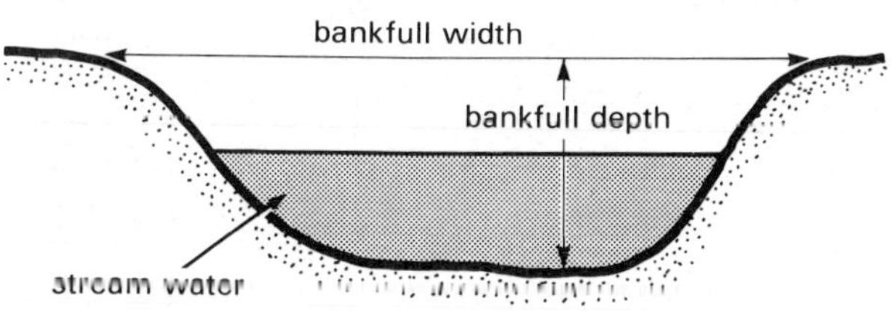

Fig 3.23 Bankfull channel dimensions

The efficiency of a channel is controlled by the degree of contact of the channel bed and banks with the stream water. Naturally with a lot of contact between water and banks (as in a wide, shallow stream) the loss of energy due to friction is very large: an inefficient channel. An indicator of channel efficiency is the 'hydraulic radius', which may be calculated from the following formula:

$$\text{hydraulic radius} = \frac{\text{CSA}}{\text{WP}}$$

where CSA = cross-sectional area
WP = wetted perimeter

Along a given channel cross-section the cross-sectional area is the area of the stream (in square metres) and the wetted perimeter is the length of bed and banks in contact with the water (in metres).

Figure 3.24 explains the terms. The value obtained for hydraulic radius is a unitless number. Higher values indicate greater efficiency.

'Width:depth ratios' of channels are a good way of describing the shape of channel cross-section. Measure width and depth as described above and divide the former by the latter. The ratio is one of several factors controlling stream efficiency. The lower the ratio, the more efficient the stream.

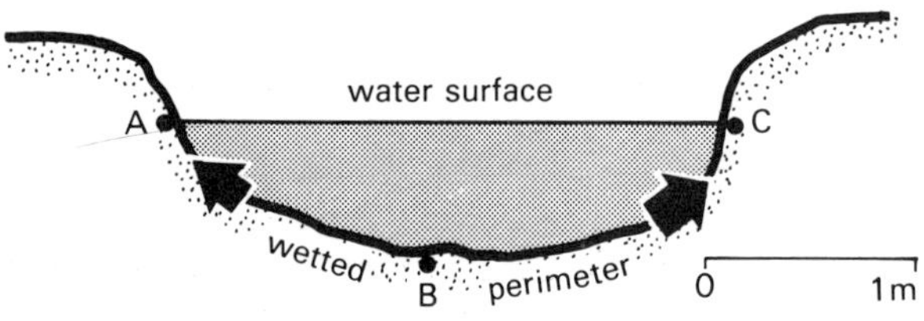

Cross-sectional area = 1 m²
Wetted perimeter (ABC) = 3 m
Hydraulic radius = $\frac{1}{3}$
= 0.33

Fig 3.24 Measurement of hydraulic radius

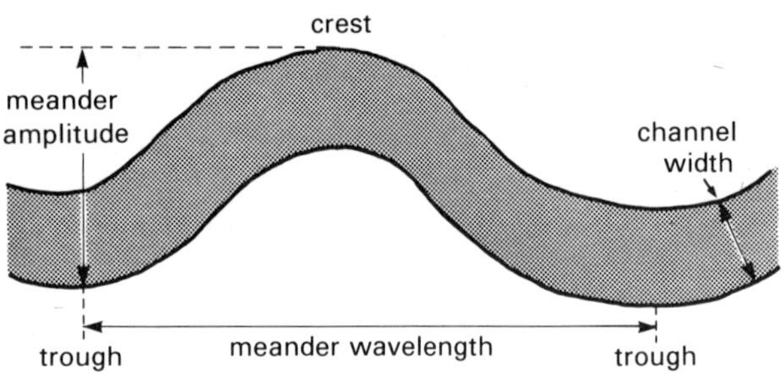

Meander wavelength: Holding a tape-measure tight, measure the straight-line distance from the trough or crest of one meander bend to the corresponding point on the next meander.

Meander amplitude: Measure the distance across the meander belt from the trough of one meander to a point aligned with the crest of an adjacent meander.

Fig 3.25 Meander geometry

Rivers exhibit many different 'plan shapes'. Meandering and braided rivers are the most striking although ideal examples are not commonly found. Meander dimensions (amplitude and wavelength) are defined in figure 3.25. These dimensions can be measured with a long tape-measure. River channels which have split into several sub-channels are known as braided channels. Because water flows in several smaller channels rather than in a single large channel, the hydraulic radius tends to be low.

River load

Material transported in rivers is called 'load'. Rivers carry three types of load: 'bed load' – boulders and pebbles that are rolled along the river bed; 'suspended load' – silt and clay particles that are carried along in the body of the stream, buoyed up by its turbulence; and 'solution load' – dissolved minerals.

The majority of solid particles carried by rivers originate outside the channel. Slope processes of wash, rainsplash and creep are responsible for carrying them into the river. Once they are in the river they may be carried thousands of kilometres and the finest particles may eventually be deposited on the sea floor.

Some of the techniques for measuring the different types of load will now be described:

Measurement of bed load Bed load is carried only at times of high discharge. If flow *is* swift then it may not be safe to wade, so take great care when collecting bed load data. A technique for measuring the amount of bed load being carried involves the use of 'bed load traps':

1 Dig a trench about 25 cm deep across the centre of the stream.
2 Insert a wooden box of approximately the same dimensions as your trench. Ensure the top of the box does not stick up above the river bed. Attach a flap to the upstream side of the box (fig. 3.26).
3 After a given interval, an hour or a day depending on the volume of bed load being transported, remove the box from the river and weigh the material.

Stream competence As the measurement of total bed load volume is rather an awkward procedure, it may be simpler to determine the 'competence' of a stream. This is the strength of a stream as reflected by the largest sized particle being moved as bed load. Stream competence may change from day to day with changes in discharge.

1 Take a dozen pebbles of similar size and paint them the same colour with waterproof paint. Take a dozen slightly larger pebbles and paint them a

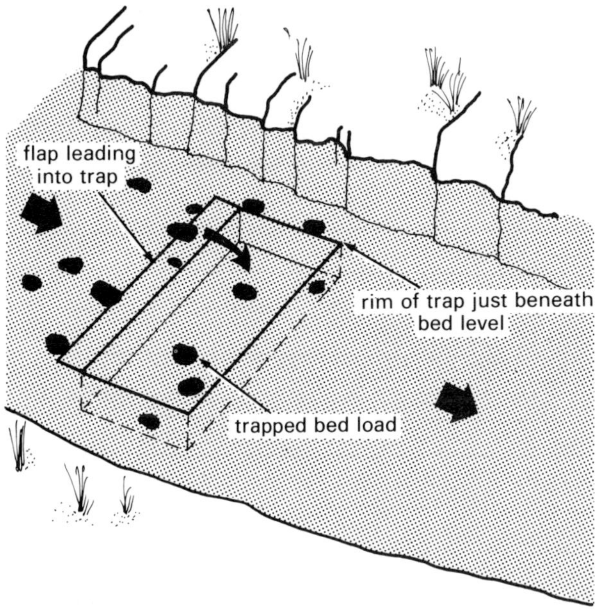

Fig 3.26 Construction of a bed load trap

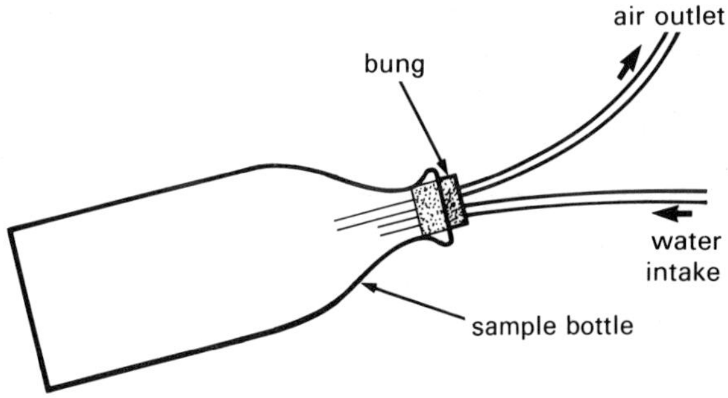

Fig 3.27 Suspended sediment sampler made from a glass sample bottle

different colour. Altogether you will need about four groups of pebbles covering a range of sizes.

2 Place the pebbles in the stream. If the stream is below bankfull stage only the smallest particles will be moved. When in flood and flowing swiftly the stream may move pebbles of all sizes.

3 Measure and record the distance the different sizes of pebbles are moved downstream. The colouring will help you recognise and distinguish the different size categories of your stones.

Measurement of suspended sediments There are several types of sampler that can be used to measure suspended load. Samplers are produced commercially but these are very expensive so it is better to make your own. You can make one from a glass flask and piping as shown in figure 3.27, or simply use a very large orange squash container. You must attach a pipe to let out air.

1 Slowly lower the sediment sampler into the stream, down to the bed and then up again. Try to judge the operation so that the sampler is just full by this time. Ensure that when taking the sample you stand downstream of the sampler or you will stir up sediments.

2 Measure and record the volume of the whole sample (in litres). Figure 3.28 is an example of a field data sheet on which to record such data.

3 If many samples are being collected in the field allow them to settle, then carefully pour off the clear water.

4 In the laboratory the sediment-laden water is then filtered through a filter paper of known dry weight. The paper can be placed in a funnel and the water left to drip through. Use Whatman 542 or 541 papers because, although slow, they do filter out even the very finest particles.

5 The filter paper plus sediment is then dried in an oven (105°C for 2 hours), removed and weighed.

6 By subtracting the known dry weight of the filter paper from the total dry weight, the weight of the sediment sample can be found.

7 As the volume of the original sample has been measured and recorded (step 2) the *concentration* of suspended sediment can be found and expressed in grams per litre (g/l).

8 If stream discharge is measured at the same time as samples are taken it is possible to calculate the *total* suspended load; simply multiply the concentration by the discharge. For example:

Suspended sediment concentration	= 0.2 g/l
Discharge	= 45 l/sec.
Total load in suspension	= 45 × 0.2
	= 9 g/sec.

This means that in one second 9 g of suspended load passed a fixed point on the bank.

Measurement of solution load In some parts of the country, particularly limestone areas, most of the load transported by rivers is material carried in solution. Most rocks are weathered by chemical action but limestones are particularly prone to this type of breakdown. Once weathering has occurred the dissolved minerals are carried away in the groundwater to a stream.

FIELD DATA SHEET

DATE: .... GROUP: ....

SITE NAME/No: .... STREAM VELOCITY: .... M/SEC

TIME: .... STREAM DISCHARGE: .... L/SEC

Sample number	Volume (litres)
1	0.80
2	0.75
3	0.85
4	0.80

LABORATORY DATA SHEET

Number of filter paper	Weight of dry filter paper (grams)	Weight of filter paper + sediment after drying (grams)	Weight of sediment (grams)	Volume of sample (litres)	Concentration (grams/litre)
1	1.05	1.30	0.25	0.80	0.31
2	1.00	1.25	0.25	0.75	0.33
3	1.10	1.30	0.20	0.85	0.24
4	1.00	1.27	0.27	0.80	0.34
....	....	....	....	....	....

Fig 3.28 Field and laboratory data sheets for the measurement of suspended sediments

A conductivity meter is a device which can measure the amount of solution load in a stream. Attached to the meter is a probe containing two electrodes. When inserted in a stream this apparatus measures the electrical conductivity of the stream water (fig. 3.29). The greater the concentration of dissolved minerals the greater the conductivity of the water.

The conductivity meter enables you to compare the *concentration* of solutes (dissolved minerals) in different streams. A stream draining a limestone catchment area would probably have a high concentration of solutes owing to that rock's susceptibility to chemical weathering. Similarly the *route* that water takes through a drainage basin can also affect the concentration of solutes. Stormflow (overland flow and throughflow), for example, tends to have a low concentration of solutes. This is because stormflow moves quickly and subsequently has little time to dissolve and transport minerals. Groundwater, on the other hand, may have spent months in contact with the rock and will have dissolved a greater volume of material. The concentration of solution load (and thus the conductivity) will therefore tend to be much greater in groundwater.

Conductivity meters can be obtained from Walden Precision Apparatus Ltd, Shire Hill, Saffron Walden, Essex, CB11 3BD, and other suppliers.

Fig 3.29 Using a conductivity meter

Project suggestions

The techniques used to measure river channel size and shape could be employed in one of the following projects:

1 Examine the relationships between channel width and depth, and the velocity and discharge along a stream. Figure 3.30 shows how such data may be displayed. Do you find such relationships hold for your stream?

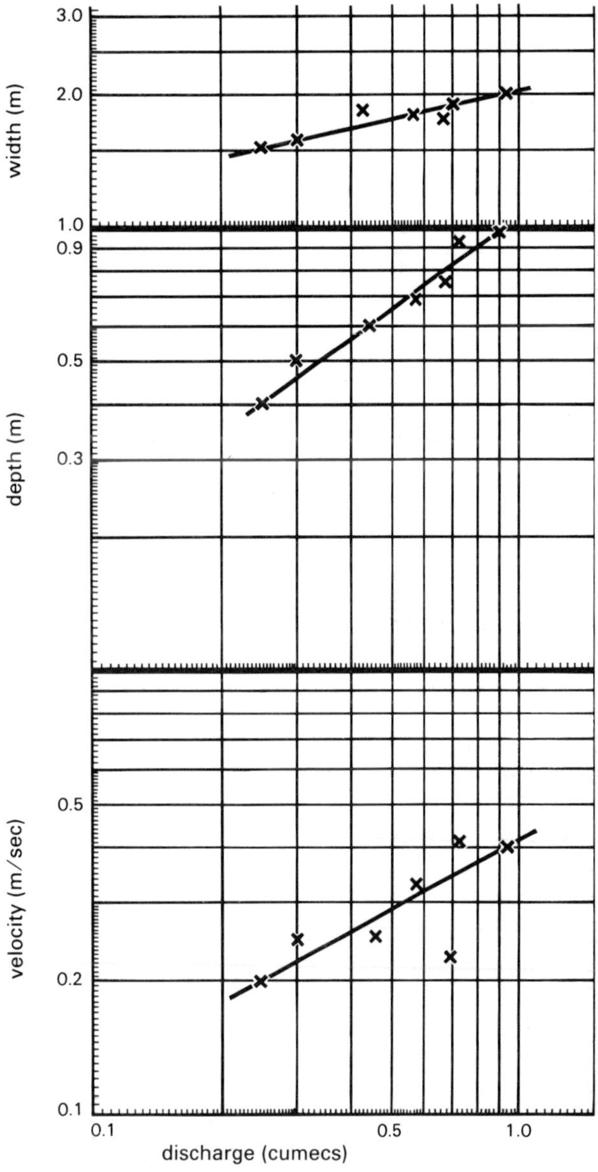

Fig 3.30 Width, depth, velocity and discharge readings taken along Ashes Brook, Church Stretton, Shropshire

2 Measure stream velocity at regular intervals along a channel. If there is much variation between one reach and another how can this be explained? Is the channel in one reach steeper, more efficient or straighter than in another?
3 Investigate meanders by comparing meander wave-length and channel width. What is the ratio between the two? Does the ratio remain constant from reach to reach or even on separate rivers? Can you suggest a reason for this relationship?

Fluctuations in river load, either in time or in a downstream direction may form the basis of a worthwhile project.

4 See whether the concentration of suspended load or the total suspended load increases with discharge. Following a severe storm concentrations of load usually rise many-fold.
5 In which sections of a stream is most bed load transported? How do discharge and stream velocity affect load? What is the critical velocity required to transport load of say 2, 5, 10 or 15 cm diameter?
6 An investigation that reveals very interesting results is the measurement (with a conductivity meter) of solute concentration before, during and after a rainstorm. If you also measure discharge and rainfall (section 3.2) and measure the conductivity of the rain water you will have a set of very interesting data to compare and relate. Alternatively, if you live in an area of mixed geology, locate, with the aid of a geological map (see page 10), streams which run on different rock types. Measure the conductivity of a stream on each rock type. Which stream has the highest concentration of solutes? Does this correspond to the most easily dissolved rock type or the most permeable rock?

3.4 Coasts

Most coastal research is concerned with investigating the behaviour of water and sediments along the shoreline, these being the processes responsible for the formation of coastal landforms. The fundamental force in the fashioning of such features is wave energy.

Wave energy

The amount of erosional and depositional activity along a shoreline depends largely on the quantity of energy arriving in the waves. Wave energy is controlled by the size and frequency of waves and is determined by the following technique:

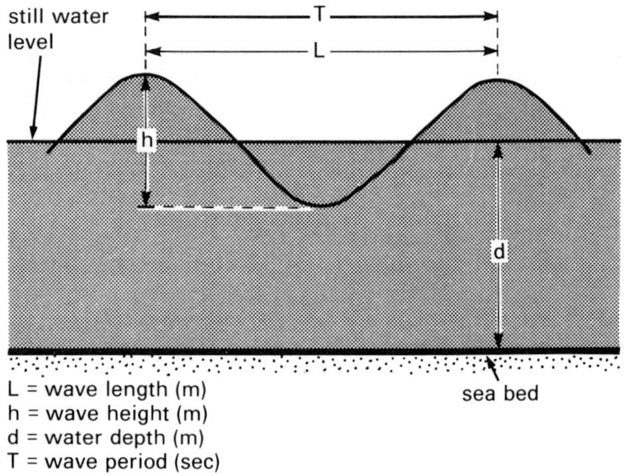

Fig 3.31 Definition of wave dimensions

1 'Wave height' (see figure 3.31 for explanation of terms) is measured (when possible) by getting in the water or standing on a groyne with a range pole (fig. 3.32). See how far up the pole wave crests reach and how low down the pole wave troughs dip. The distance between the two is the wave height. Alternatively you may be able to observe waves passing a fixed obstacle such as a pier leg or groyne. In this case estimate the average height of the waves against the object.
2 'Wave length' is more difficult to assess visually. Fortunately wave length is related to 'wave period' (the time between two wave crests passing the same point). Time fifty waves passing a fixed point and calculate the average wave period. Wave length can then be found from one of the following equations:
For deep water waves (water depth $>\frac{1}{2}$ wave length):

$$L = 1.56 \times T^2$$

where L = wave length in metres
T = wave period in seconds

For shallow water waves (water depth $<\frac{1}{2}$ wave length):

$$L = 3.13 \times T \times \sqrt{d}$$

where d = depth of water in metres
3 Wave energy can then be calculated:

$$E = 740 \times h^2 \times L$$

where E = wave energy in joules per metre (J/m) width
h = wave height in metres
L = wave length in metres

The energy of the waves will vary enormously, depending on sea conditions. On a calm day with wave heights of around 0.5 m, each metre width of

Fig 3.32 Measuring wave height – a life jacket would have been a good idea. . . .

wave will be transferring about 1000 J of energy to the beach. Following a storm when waves are 3 m high the energy expended by them will be over 1 000 000 J/m width.

4 'Wave steepness' can also now be calculated:

$$\text{wave steepness} = \frac{\text{wave height}}{\text{wave length}}$$

Values for wave height and wave length must however be collected in deep water (i.e. water depth $> \frac{1}{2}$ wave length). Wave steepness is a critical factor in the shaping of beach profiles.

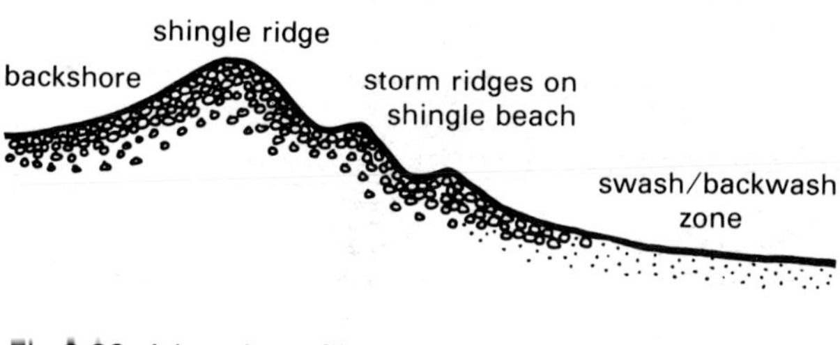

Fig 3.33 A beach profile

Beach profiles

A beach profile or cross-section (fig. 3.33) reflects several factors, notably the size of beach material and the steepness and type of wave that modelled it. Generally beaches composed of shingle can support quite steep gradients, around 1:4. Chesil Beach at Abbotsbury in Dorset maintains gradients between 1:2 and 1:3. Fine sandy beaches support gradients not usually in excess of 1:30. Larger beach sediments are more stable and have higher angles of repose. Percolation rates on beaches may be measured using the technique for measuring soil infiltration rates, described on page 16.

There are two main types of wave that break along shorelines, 'spilling breakers' and 'plunging breakers'. Figure 3.34 illustrates the two wave types. Spilling breakers have a long upbeach rush (swash) but little return flow (backwash) and are associated with gentle beach gradients. Plunging breakers, however, exert a strong downwards force as they break and carry material seawards with their strong backwash. These waves are associated with steeper beach profiles.

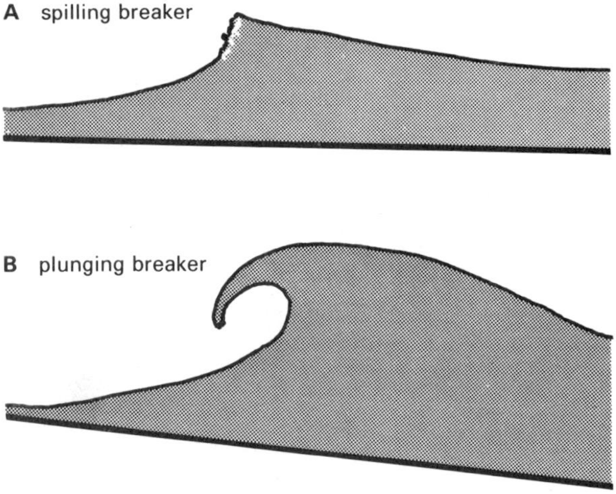

Fig 3.34 Wave profiles

The method for profiling on beaches is exactly the same as described on page 11. With a clinometer, two range poles and measuring tape, record the gradient and distance between each break of slope. Mark onto your profile particle size (page 31) and features such as storm beaches.

Longshore drift

Waves often arrive oblique to the beach. This causes a lateral shift of material known as longshore drift. The details of the mechanism are explained in figure 3.36. The techniques for measuring the rate of drift are outlined here:

1 Marker pebbles:

a Spray paint an assorted sample of at least two dozen pebbles of different sizes and leave to dry.

b Select a clear stretch of beach away from any groynes that will interfere with longshore drift. Put the stones in the swash/backwash zone and stick a pole firmly in the ground to mark their position.

c After a count of fifty waves locate and plot the direction and distance the stones have travelled. Some stones will have vanished and some will possibly move in the opposite direction to the majority, but hopefully the bulk of your pebbles will give you an indication of the direction and rate of drift.

d This experiment tells you the direction of longshore drift at a particular moment in time. However, because the direction of drift varies with

Fig 3.35 Measuring the height of beach material against a groyne as an indicator of longshore drift

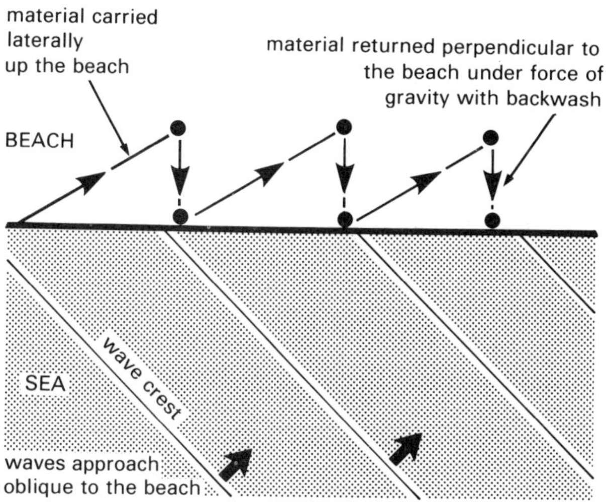

Fig 3.36 The mechanism of longshore drift

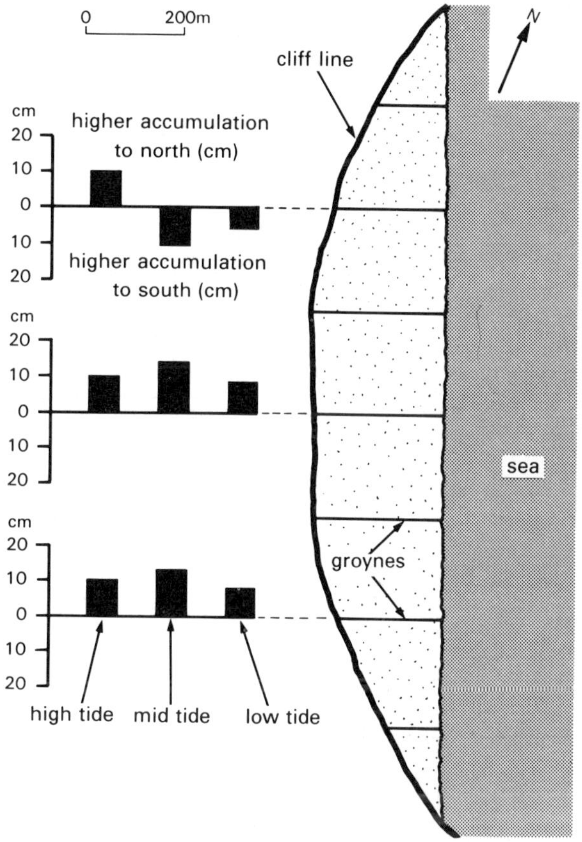

Fig 3.37 Sketch map showing accumulation of sediments against groynes (on Swanage beach, Dorset) as an indicator of long-term longshore drift activity. Values refer to the difference in beach height on either side of each groyne.

changes in prevailing winds, you will get a better picture of material transfers by running your experiment over a longer period. In this case paint several hundred pebbles of varying sizes and trace their movement over a period of days or weeks. Record the different rates of movement of large and small pebbles, making a note of weather and wave conditions at the same time.

2 Groyne measurements A picture of longer term pebble movements may be derived using the following technique:

a Measure the difference in the height of beach material on both sides of a groyne (fig. 3.35).
b Make these observations at the high tide mark, at the mid tide mark and as low as possible down the groyne at low tide.
c Plot your results on a map (fig. 3.37).

Beach sediment analysis

Geographers can learn much about the processes that fashion coastal features by examining the size and shape of particles along the shoreline. Even a fairly rudimentary analysis of particle size and shape has considerable practical value. For example, the roundness of a pebble may indicate the amount of attrition (wearing) to which it has been subjected. Freshly shattered fragments derived from cliffs are sharp and angular. After prolonged wearing by waves they become more rounded and smaller.

Measurement of particle size Sediments of about 4 mm radius and over can be easily and accurately measured in the field with either a ruler and calipers or with a tape-measure. Measure the longest axis to find particle size. Having measured each particle in your sample, establish suitable size class-intervals (0–19 mm, 20–39 mm, 40–59 mm, etc.) and plot a bar graph. You can then make a visual comparison of particle size distribution along a beach.

Measurement of particle shape:

1 Cailleux roundness index:

a Take one pebble from your sample. Measure and record the long axis (*a*).
b Holding the pebble flat, lay the sharpest corner on a chart of concentric semi-circles of known

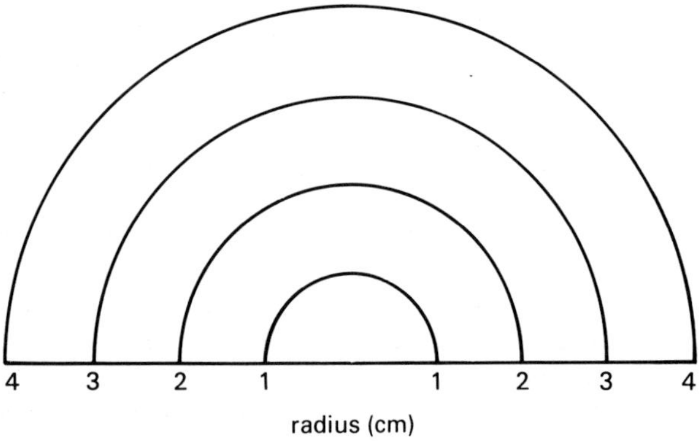

Fig 3.38 Concentric rings for measuring pebble radius for Cailleux roundness analysis. Place the sharpest corner of the stone (in its flattest plane) on the chart to assess the radius of curvature.

radii (fig. 3.38) and assess the radius of this corner (*r*). See figure 3.39.

c Repeat for each pebble in your sample then substitute your figures in the following equation:

$$R = \frac{2r}{a} \times 1000$$

where R = Cailleux roundness index

d Calculate the mean R value for the sample. (See page 91 for calculation of mean values.) Values of R will lie between 0 and 1000, 1000 representing a perfectly circular stone.

The higher the figure obtained the more rounded the pebble. High values will be indicative of exposure to wearing for a very long time or of very active attrition. Pebble shape will of course also reflect lithology. Very hard rocks such as flints and granites, or rocks with cleavage planes which tend to flake, would probably never match the almost perfect sphericity sometimes found in chalk. When studying sediment shape it is therefore better to confine your measurements to pebbles of the same rock type.

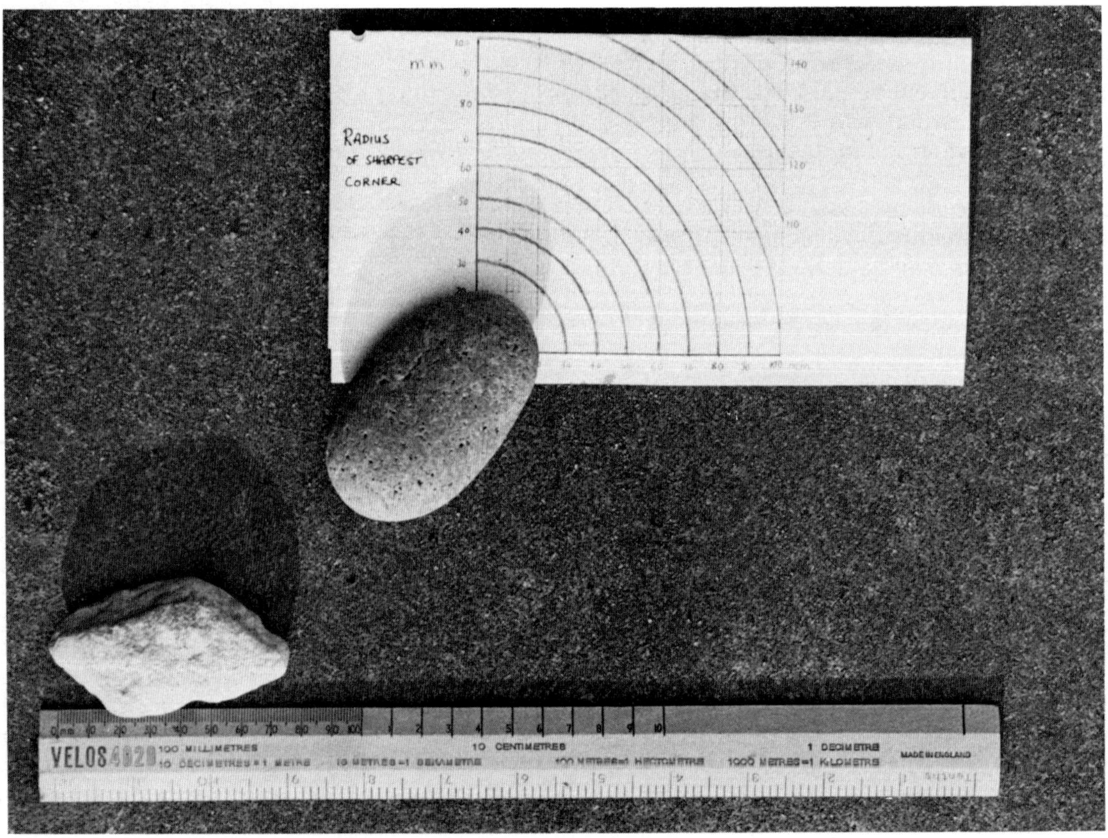

Fig 3.39 Measuring long axis and radius of the sharpest corner for Cailleux roundness analysis

2 Powers' scale of roundness For a speedier, but less accurate technique, Powers' visual comparison chart is a valuable aid (fig. 3.40). Compare each pebble in your sample with Powers' chart and note the number of particles in each category. A bar graph may be drawn up and a comparison made between several samples taken at different points along a beach.

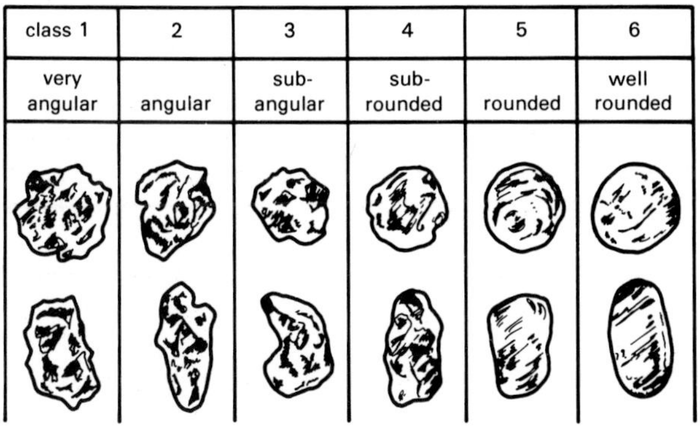

Fig 3.40 Visual chart for Powers' scale of roundness

Cliffs

Information on cliffs is required to enable us to compare and relate their height and shape with geological factors and with intensity of wave attack.

Cliff height is estimated visually, or better still determined by elementary surveying (fig. 3.41). Cliff slope is ascertained with a clinometer (page 22) and its geology identified, if necessary, with the aid of a local guidebook or geological map (page 10).

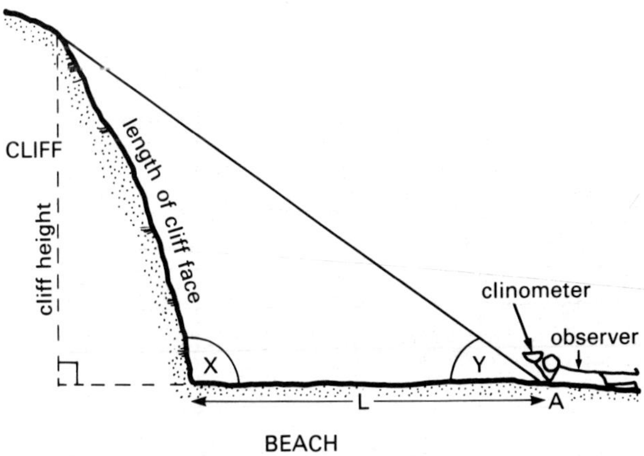

Fig 3.41 The technique for surveying cliff height. Measure angle X formed by cliff and beach, and distance L from cliff foot to a point (A) at which the angle (Y) formed with the cliff top can be measured. Cliff height can then be determined by producing a scale drawing.

Broadly speaking, questions concerning cliff steepness and shape can be answered by a number of observations. How densely are the rocks jointed? How hard are they – crumbly like chalk or soft like clay? Does mass movement take place – sliding or rotational slumping? What degree of protection is afforded by the beach or by man-made defences? Do waves break against the cliffs at high tide? Is there an accumulation of freshly weathered material at the foot of the cliff or is it promptly removed by the sea? Is the wave attack usually powerful or are the waves only small and incapable of vigorous erosion? Are the cliffs vegetated – if so, how densely and with what? What is the history of sea-level changes in the locality?

Project suggestions

Beaches are particularly interesting areas to study due to the speed at which change occurs.

1 An interesting project is the comparison of a beach in summer with the same beach in winter. Record wave energy, wave type, beach profiles, and particle size during the low energy conditions of summer. Make these observations again at the same site following stormy winter weather and compare the two. How much more powerful were the storm waves? How has the profile of the beach changed? Is this due to the presence of predominantly larger particles or the action of different types of wave?

2 Some stretches of coast have markedly different conditions even over quite short distances. Many bays (those along the South Cornwall coast for instance) have headlands which absorb most of the energy unleashed by Atlantic breakers, giving very sheltered bay-head beaches behind the headlands. Such an area would be suitable for a detailed comparison of wave energy, cliff steepness, beach profiles and composition.

3 A similar project is to relate the combined influence of geology and marine action on cliff steepness in an area underlain by rocks of widely different natures. For example in the limestone, chalk and clays of the South Dorset coast a simple 'compare and contrast' type exercise is possible.

4 Sediment analysis helps us understand the formation of features such as spits. Measure particle size and shape at regular intervals along a spit. Measure rates of longshore drift at the same sites, noting relative speeds of large and small particles. Longshore drift itself can be correlated with predominant wind directions and strengths. Longer term wind data could possibly be obtained from the local coastguard.

3.5 Soils

Soil is composed of weathered rock particles, decayed vegetation, water and air (fig. 3.42). The nature of soil is strongly influenced over time by climate. Figure 3.43 summarises the factors that together affect the development of soil.

This section explains some of the techniques used to investigate the composition, texture, physical and chemical properties of soil. Although these soil characteristics are of interest in their own right, suggestions for projects which relate soil qualities to other environmental phenomena (such as vegetation and microclimates) are made at the end of this chapter.

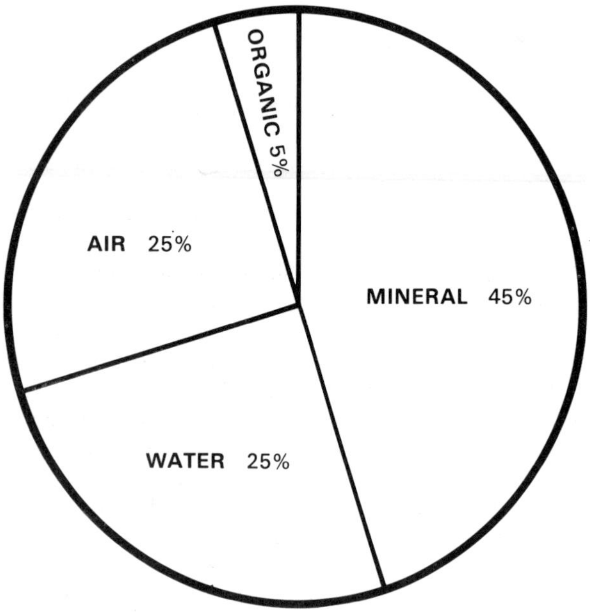

Fig 3.42 Volume composition of a typical UK topsoil

Soil sampling

Before you can begin analysing soil quality you will need a number of carefully collected samples. If you are studying the changes occurring at different depths in the soil you will need to collect your samples from a 'soil pit'. This is a pit in the ground, preferably dug down to bed rock, one side of which is cut straight and cleaned. This side of the pit will reveal a 'soil profile', a series of layers in the soil. These layers or 'horizons' are characterised by differences in colour, composition and texture. The upper part of the soil profile is rich in organic matter, added to the soil as

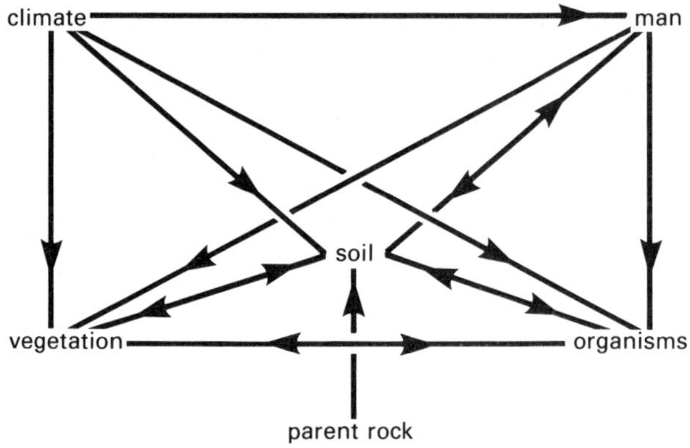

Fig 3.43 A diagrammatic illustration of some of the interrelationships between soil forming factors

vegetation dies and decomposes; the lower soil horizons are more stony, due to the proximity of the bed rock. In an undisturbed soil (i.e. unploughed) it may prove possible to identify several distinct horizons. These layers are conventionally designated by certain letters to aid comparison of soil types. Figure 3.44 illustrates and explains a typical UK soil profile.

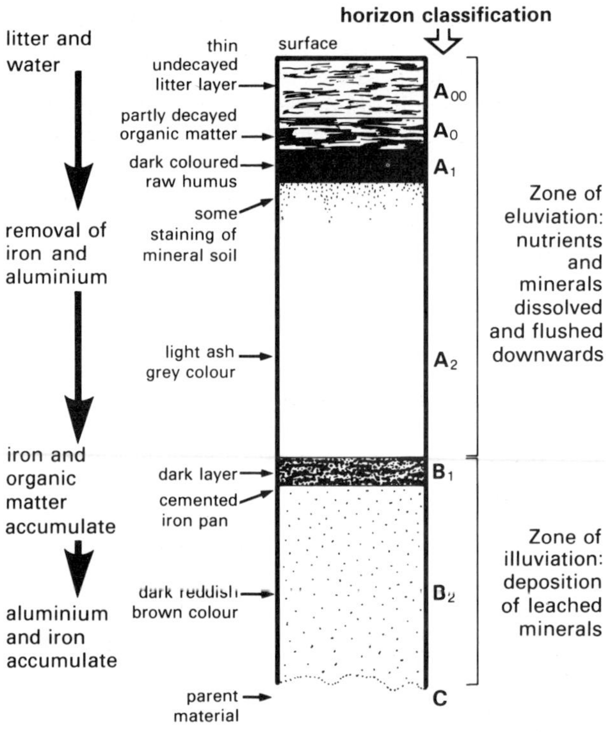

Fig 3.44 The development of a typical UK soil profile (a podzol). This occurs where there is a net downward movement of water; precipitation is greater than evaporation and soil texture allows free drainage.

Soil pits are only needed if studying soil profiles. When investigating spatial variations in soil quality simply dig aside the vegetation and humus and take your sample from the top of the 'true' soil.

Take a supply of tough polythene bags in which to collect the samples. Put each sample into a polythene bag along with a numbered label and tie the bag tightly to prevent water loss. At the same time, record in your field log-book the date and time of collection, the location and number of the sample, the depth from which the sample was taken, local vegetation type and weather conditions.

Soil composition

Soil water content

Soil water is important in many ways. It provides plants with moisture and acts as a medium of transport for nutrients. Very dry soils are agriculturally poor since crops cannot obtain sufficient moisture and consequently they wilt. However, waterlogged soils are also poor. Water occupies nearly all the pore space, thereby excluding air from the ground. Many micro-organisms that are essential for nutrient release live on oxygen. They cannot survive under waterlogged conditions and the supply of nutrients to plants is therefore reduced. The method for measuring soil water content is as follows:

1 Crumble a small amount of soil (around 30 g) into a small pre-weighed crucible and weigh crucible plus soil.
2 Put the sample in an oven at 110°C overnight to remove the water.
3 Take the sample from the oven and weigh the crucible and dry soil together.
4 Subtract the weight of the crucible to give the dry soil weight.
5 Subtract the weight of the dry soil from the weight of the wet soil to give weight loss on drying (total weight of water evaporated from the soil).
6 Substitute your figures in the following equation:

$$\frac{\text{weight loss on drying}}{\text{weight of wet soil}} \times 100 = \text{percentage of water (by weight) contained in the soil}$$

Soil water content will vary from day to day with weather changes, from place to place (for example, down a slope), from soil type to soil type and with position in the soil profile.

Soil organic content

When plants die and are left to rot, bacteria rapidly bring about their decay. The nutrients the dead plants once took from the soil are returned to the ground. If the soil is to remain fertile then organic matter in the soil is clearly essential. However, as we shall shortly see, very high concentrations of organic matter do not necessarily correspond with highly fertile soils.

To determine the weight of organic matter in a soil sample:

1 Take the dry soil sample from the previous experiment and grind it with a mortar and pestle.
2 Return the ground, dry soil to the crucible and weigh it.
3 Heat the soil sample and crucible over a bunsen burner for half an hour. This should ensure the incineration of all organic matter.
4 Wipe off any carbon deposited on the underside of the crucible. Weigh the burned soil plus crucible and subtract the known weight of the crucible.
5 Subtract the weight of the soil after burning from the weight of the soil before burning to give the weight loss on burning (i.e. the weight of organic matter).
6 Substitute your figures in the following equation:

$$\frac{\text{weight loss on burning}}{\text{weight of dry soil}} \times 100 = \text{percentage of organic matter by weight of dry soil}$$

The concentration of organic matter in soil varies considerably. In arable land values within the topsoil lie between 1% and 5%. Peats, often of very low fertility, contain up to 80% organic matter. The actual volume depends on the balance between the rate of gain and the rate of loss of matter. Certain vegetation, for instance deciduous woodland, supplies large amounts of organic matter whereas root crops supply very little. Loss of organic matter depends largely on the rate at which it is broken down. Waterlogged soils with low oxygen content contain few organisms; the breakdown of dead vegetation is therefore very slow and the concentration of organic matter may become quite high – as with peats. Well drained, aerated soils allow a more rapid breakdown of dead vegetation and a more speedy return of nutrients to the soil.

Soil chemistry

Having established the percentage weights of the soil components it is useful to know something about the chemical properties of the soil. The chemical composition of soil is determined by the type of parent rock, vegetation type, rainfall totals and average temperatures. The single most important chemical characteristic of soil is its acidity or alkalinity as indicated by its pH value. The pH value of a soil is a measure of the concentration of hydrogen ions in the soil water. The greater the concentration of hydrogen ions the more acidic the soil is. In most soils pH values lie between 5 and 9. A pH value of less than 7 is acidic, pH greater than 7 is alkaline, and pH 7 is neutral.

Acidic soils are often found in areas of high rainfall. The predominant downward movement of water through the soil leaches or flushes salts out, leaving a high concentration of hydrogen ions. Alkaline soils are found where the bed rock weathers to produce salts, especially calcium, magnesium and sodium carbonates. Calcareous rocks such as limestones therefore tend to produce alkaline soils due to the availability of calcium carbonate.

The pH value of the soil does not *directly* affect crop growth except in cases of extreme acidity or alkalinity. Instead it is important because it affects the availability of nutrients for plant growth. In alkaline soils some nutrients essential for crop health (e.g. iron and magnesium) become scarce at pH values of 7.5 and over. At the opposite end of the scale, in acid soils below pH 5.0, the over-abundance of nutrients such as aluminium and iron, mobilised by the acidity, may prove toxic to crops. Some plant species prefer acidic, and others alkaline soils. Among the former are gorse, potatoes, rhododendrons and heather. Examples of alkaline-tolerant species include lettuce, sugar beet and cauliflowers.

Determination of pH: Colorimetric method Of the techniques available for the determination of soil pH the most accurate and appropriate is the colorimetric method. The chemicals needed for the test (fig. 3.45) may be obtained from British Drug House Chemicals Ltd, Poole, Dorset, BH12 4NN, or from large gardening centres. To determine pH:

1 Pour a small sample of soil into a test tube.
2 Add barium sulphate, distilled water and the indicator liquid.
3 Shake the tube thoroughly and leave to stand until a clear coloured fluid develops.
4 Compare the colour of the fluid with the chart provided and read off the pH to the nearest 0.5 of a unit.

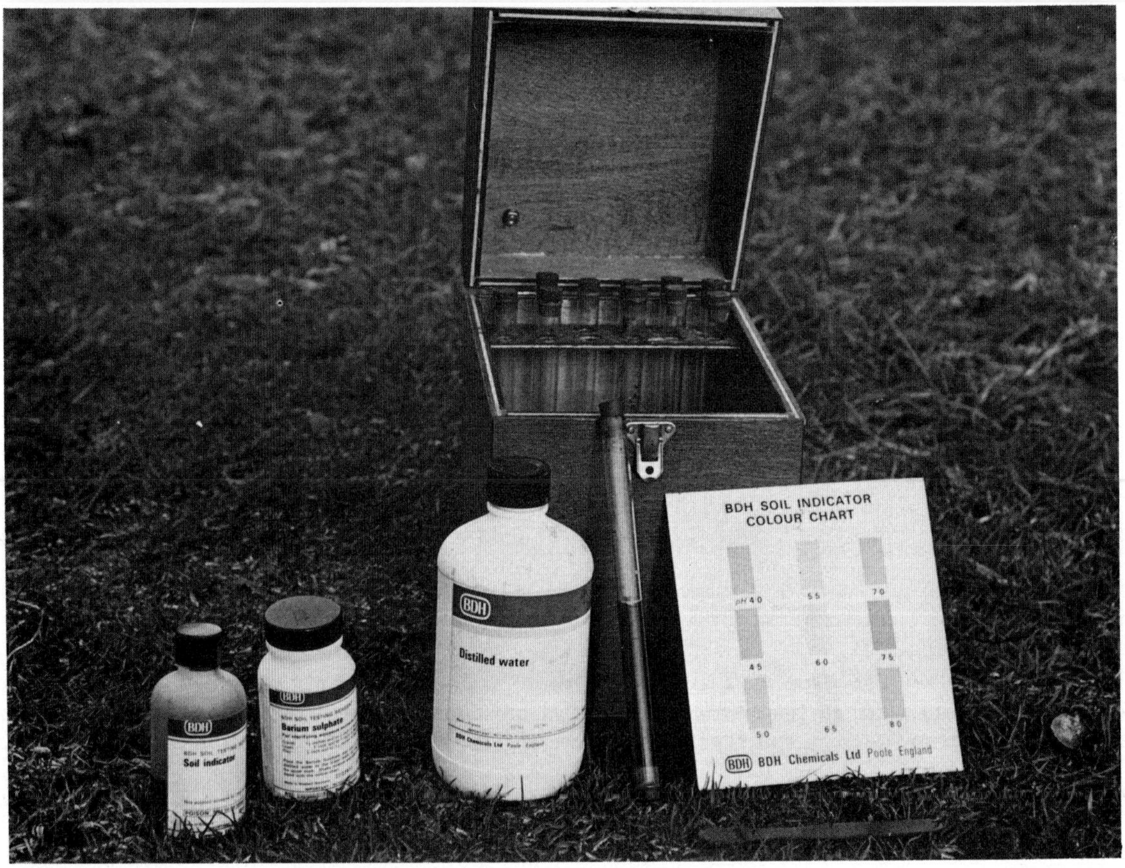

Fig 3.45 The BDH kit for testing soil pH

Soil texture

Soil texture refers to the size of the particles that make up the soil. There are three broad categories of soil which you will be able to identify: sands, loams and clays. Sandy soils are light, well drained and well aerated, and are composed predominantly of fairly coarse grains (less than 2 mm diameter). Clay soils are sticky, often waterlogged, and consist largely of particles less than 2 μm (2×10^{-3}mm) diameter. Loamy soils are a combination of the two, sand and clay; these make excellent agricultural soils, being both well drained and fertile. Obviously not all soils fit neatly into these three categories. A number of intermediate soil classifications exist, such as sandy loams and clay loams.

The simplest and quickest way of assessing soil texture is to roll a wetted sample in your hand. This is called 'hand texturing':

1 Take a handful of soil and dampen it slightly.
2 Rub it between your fingers, feeling for grainy particles and cohesion.
3 Refer to figure 3.46 to classify.

Soil texture	Soil characteristics when wet
sand	feels gritty; not sticky even when wet
sandy loam	may be moulded into a thread with difficulty; not sticky
loam	does not feel sticky but will roll into a thread
clay loam	sticky; will easily form a thread
clay	very sticky; can be moulded into any shape

Fig 3.46 Table for hand texturing of soil

Soil bulk density

Bulk density is the weight of a soil sample divided by its volume expressed in g/cm³. The bulk density of a soil sample reflects two things: the proportions of mineral and organic matter and the degree of compaction of the ground. As mineral matter is very dense (2.5 g/cm³) compared with organic matter (0.5 g/cm³) mineral soils with a high proportion of weathered rock particles have a greater density than organic soils. Bulk densities of samples taken in a soil pit will increase with depth due to decreasing proportions of organic matter, less root penetration and compaction by the overlying soil. The method for measuring bulk density is as follows:

1 To collect a sample for analysis you will need a soil corer or auger. This is a small metal cylinder, open at one end, which is hammered about 6 cm into the ground. If you cannot obtain a soil auger you can do just as well with a short length of sturdy plastic piping, about 3–4 cm in diameter.
2 Remove the corer and from the hole left in the ground measure the depth of the soil core obtained. The core itself will have been compressed and would give a false reading if measured.
3 Measure the radius of the soil core.
4 The volume may be found from the following equation:

$$\text{volume} = \pi r^2 h \ (\text{cm}^3)$$
$$\text{where } r = \text{radius (cm)}$$
$$h = \text{depth (cm)}$$
$$\pi = 3.14$$

5 Having found the volume weigh the sample. Substitute your figures in the equation below:

$$\text{bulk density} = \frac{\text{weight (g)}}{\text{volume (cm}^3)}$$

The average bulk density of a mineral soil is around 1.25 g/cm³ and of peat around 0.5 g/cm³.

Soil temperature

The temperature of a soil is important for the germination and growth of seeds and plants. Below certain critical temperatures germination and root growth will not take place. Corn seed for instance only begins to germinate once the soil temperature rises above 7–10°C and reaches optimum growth around 35°C.

Soils obviously take longer to warm and cool than the air. Different soils heat up and conduct warmth more rapidly than others (i.e. they have a greater thermal conductivity). Water content is the single most important soil characteristic controlling the rate at which a soil will heat up. A badly drained clay soil will tend to be much cooler than a well drained sandy soil as the greater rate of evaporation from the former will cause more heat loss (due to the latent heat of evaporation).

Setting up a site to investigate soil temperature is quite straightforward:

1 Dig a soil pit 50 cm deep.
2 Insert a thermometer horizontally at least 15 cm into one side of the pit at 2 cm depth. Add two further thermometers at depths of 10 cm and 30 cm.
3 Cover the pit with a lid to keep out the rays of the sun.
4 Take temperature readings at four-hourly intervals over a 24-hour period – as late at night and as early before sunrise as possible! Record the air temperature at the same time as you record the soil temperature.

5 Plot the readings of the three thermometers on one graph and observe the lag behind the air temperature that each displays.

Project suggestions

Many soil properties are interrelated: for example, bulk density and soil composition; soil moisture content and soil texture. You can investigate the extent to which they are related, either in a soil profile or in soils being used for different purposes.

1 Take four or five samples across a dirt track, one sample just off the track where the vegetation is untrampled, another sample from the side of the path where traffic is light and another sample from the centre of the path which is most heavily used. Take two more samples from corresponding points on the other side of the path. Measure the bulk density, water and organic content of each sample. Is the bulk density lowest where the organic content of the soil is greatest? What is the relationship between volume of traffic on different parts of the path and bulk density? Note the condition of the vegetation (if any) at each site. If the plants are in a poor condition is this the result of trampling on them or because of poor soil quality? With a carefully prepared programme of experiments you can answer these and any other questions that your findings lead you to.

2 With a farmer's permission you could investigate the composition, pH and hydrological characteristics of his soil. If in previous years the farmer has planted several fields with the same crop he may be able to give you the crop yields. If so, analyse the soils of those fields. Does one field seem to have a markedly superior soil to another? What light does your soil data throw on the crop yields? Remember other factors will affect crop yields too (fertilisers and microclimate, for example), so try and find out if one field is more fortunate in these respects than another.

3 The study of soil temperatures can make an interesting project. Depending on availability of time you could conduct a long-term study, taking daily readings (at a standard time each day) over a period of several months, watching the soil heat up in spring or cool down in autumn. Alternatively you could compare different types of soil, e.g. sandy soils and clay. How does soil moisture content relate to the thermal conductivity of the soil? Compare sites under different types of vegetation: grass and trees. How does rainfall or snow cover affect soil temperatures?

4 Make a detailed analysis of a soil profile. Dig a pit in an undisturbed area and take samples from each horizon. Measure bulk density, organic matter and water content, pH and soil texture. Use the field data sheet in figure 3.47. Do the variations from one horizon to another match your expectations? If not, what is your explanation for your observations? Compare soil profiles taken in different areas: down a slope, on different parent rocks or under different vegetation covers.

PROFILE LOCATION: .... DATE: ....

PROFILE NUMBER:.......................... TIME:.......................................

WEATHER CONDITIONS:.................... VEGETATION COVER: ....

Sample depth (cm)	pH	Bulk density	Texture	Loss on drying (%)	Loss on ignition (%)
1					
2					
3					
4					
5					

Fig 3.47 Field sheet for recording soil profile data

3.6 Vegetation

The interest of vegetation to most geographers lies in the way plants adapt and respond to their environment. Microclimate, altitude, aspect, soil type and the activities of men and animals all have a discernible influence over the type and condition of vegetation. The techniques described in this section are chiefly concerned with the sampling and analysis of plant communities with a view to relating them with the above factors.

Vegetation sampling

In many biogeographical studies you must first sample vegetation in the area under consideration to find out which species exist in the area, their distribution and density. Two sampling methods can be used: quadrat sampling and transects. Vegetation type will determine the appropriate technique – transects for trees and quadrats for herbaceous plants.

The size of the sample must be carefully considered. Clearly an undersized sample involves the possibility that even a fairly abundant species remains unrecorded. Conversely, although an oversized sample may ensure enumeration of even very sparsely represented plants, the effort put into the sampling will not be matched by the results. Therefore you must be alert to the degree to which your samples are representative.

Theoretically sampling should be random, although such an approach may result in large areas being left unsampled and the exclusion from the sample of an important species. This problem may, to a large extent, be overcome by following a systematic sampling network (e.g. divide a wood into grid squares) and then sampling at random within each particular vicinity (see chapter 2).

1 Quadrat sampling:
A quadrat is a frame, usually one metre square, which is thrown 'blind' in the sampling area. The percentage of the frame occupied by each species is estimated, together with the percentage of bare ground. For sampling small shrubs and bushes you may need a slightly larger quadrat. In this case it is not necessary to use a quadrat frame. Simply mark out an area of say three or four metres square and sample within this. This quadrat can still be selected randomly. For instance you can throw a stick and take the point where it lands to be the north-west corner of your quadrat.

2 Transects:
A position in a wood, chosen either at random or dictated by the sampling network, is taken as the starting point for the transect. A 30 m tape is run out in a random direction. Every metre note the species of tree closest to the metre mark. Count a tree only *once* if it falls exactly between two adjacent metre markings.

Identification

For those not familiar with the flora of the British countryside a number of guides to identification exist. For example, the series of 'Observer' books is very useful for identifying trees, shrubs, flowering plants and grasses.

Vegetation condition

Studies of the response of plants to microclimate (phenological studies) clearly require a measurement of the state of vegetation. Figure 3.48 is an example of a field data sheet that includes most of the observations required in making such an assessment.

The condition or health of plants does not of course depend only upon the season and climate. Damage to plants caused by man's activity is common – along and beside footpaths, for instance. Any study of the impact of man would clearly need an assessment of the extent of damage. Figure 3.49 gives an idea of the type of observation which will reveal the degree of plant destruction.

Canopy depth and density

Some of the studies suggested at the end of this section require a measurement of the density of a wood in order to relate it to factors such as micro-climate. For example, the dragging effect of trees on wind speed is related both to tree height and the density of the tree canopy. *Tree height* can either be estimated, or determined using the technique described on page 33. The top and bottom of the tree canopy and the height of the other vegetation layers – the shrub layer, the field layer (grasses) and the ground layer (mosses, etc.), if present (fig. 3.50), can be measured and plotted in the form of a bar graph.

Temperatures are dependent largely upon the amount of solar energy received by a given area. In woodland this is directly related to the *density of the tree canopy*. Daytime temperatures in a wood are often lower than those of the surrounding area due to shading by leaves. Figures for the degree of shading may be obtained using a camera light meter. Many cameras have a built-in light meter that indicates (with a needle or by a light) when the correct aperture and shutter speed have been set. Keep the aperture (f number) constant, at say f4 or f5.6. Hold the camera horizontally, pointing out towards the edge of the wood, and record the shutter speed that *would* produce a correctly exposed picture at each site. If, in a heavily shaded area, it is $\frac{1}{15}$ second and in a lightly wooded area $\frac{1}{60}$ second then there

DATE:	SLOPE ASPECT: ....
TRANSECT NUMBER:	SLOPE ANGLE (DEGREES): ....

Flower/tree species	Height	Girth (chest level)	Leaf condition (estimated %s)						Flower condition (estimated %s)					
			bud	bursting	just out	full out	falling	bare	bud	bursting	just out	full out	falling	bare

Fig 3.48 Field sheet for assessing the condition of trees and flowers

DATE:	QUADRAT NUMBER: ....
LOCATION:	QUADRAT SIZE: ....

Coverage of most abundant species: type % cover	Coverage of second most abundant species: type % cover	Bare earth: % cover
Coverage of least abundant species: type % cover	% of plants with broken stems:	% of bushes with broken twigs:
Diameter of largest broken twig:	Comments: e.g. on possible cause of damage – walkers, vehicles etc.	

Fig 3.49 Field sheet for assessing the extent of plant damage

Fig 3.50 The structure of an English deciduous woodland

is about four times more light available to vegetation at the latter site. It is important to conduct this experiment on a day when there is the same amount of light available from minute to minute, i.e. when there are either no clouds in the sky or on an overcast day.

Project suggestions

Vegetation studies are interesting to conduct because of the sensitive and complex relationships plants have with the environment. Other sections in this book deal with techniques in micro-meteorology, soils and hydrology, and all these aspects of the natural world combine to influence the type and condition of vegetation.

1 The influence of microclimate on vegetation is particularly pronounced in early spring. An east–west oriented valley with wooded slopes provides plentiful opportunities for the investigation of this relationship. Simultaneous observations of soil temperature, air temperature, light intensity and wind speed should be made on the north and south slopes. Take these readings early in the morning, at lunch time and late in the afternoon over a period of four or five days. Take several transects through the woods on both sides of the valley and note plant condition (fig. 3.48). The climates of the two sides may be sufficiently dissimilar to produce differences in the condition of vegetation (for instance, in the number of trees in bud or full leaf). A test of significance (page 98) might be employed to see whether the difference in the number of trees in bud is truly significant or if it could be put down to chance. The results may then be related to the climatic data you have collected.

2 The effect of vegetation on microclimate can be similarly assessed. Establish three or four meteorological stations along a transect from the margin to the centre of a wood. Record precipitation, temperature, humidity and wind speed. Using the techniques described in this chapter measure the height and density of the canopy at intervals along the transect. Analyse how climate is modified along its length. Naturally the microclimate of the wood will vary with changes in synoptic weather conditions. During a period of strong winds for instance, the temperature difference between centre and periphery may disappear. Similarly in deciduous woods there will be notable contrasts between summer and winter with changes in the density of foliage.

3 The effects of man's activity on vegetation can be gauged quite successfully. Pedestrian counts along footpaths are relatively straightforward and it should be possible to identify the most heavily utilised paths within a woodland reserve. During summer, when trampling is at its greatest, the least resilient ground cover may disappear. Quadrat sampling will reveal the extent of bare ground and the coverage of different species. Subsequent sampling in the spring (when trampling is considerably less) may reveal a greater coverage of the more delicate plants and a reduction in the percentage of bare earth.

4 A vegetation project with an historical angle makes use of the fact that the number of shrub species within a hedge is a function of the age of the hedge. Observations in many English counties have shown that (with a high degree of certainty) one new species colonises the hedgerow every one hundred years. Saxon hedges have about ten shrub species, Tudor hedges four and Enclosure Act hedges two. You need to count the number of species in a 30 m stretch of hedgerow to establish its approximate age. Using this technique you could, for example, date the hedges in an old village and (with the aid of historical records and maps) see how field size and shape have changed over time.

3.7 Climate and meteorology

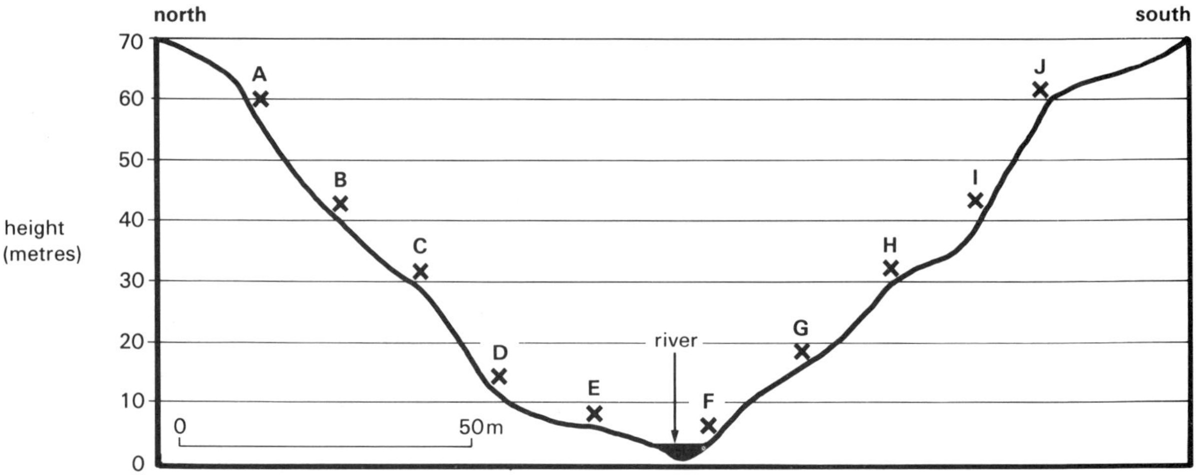

Fig 3.51 (a) Transect across the upper valley of the River Ebble, Wiltshire, showing points at which temperature readings were made

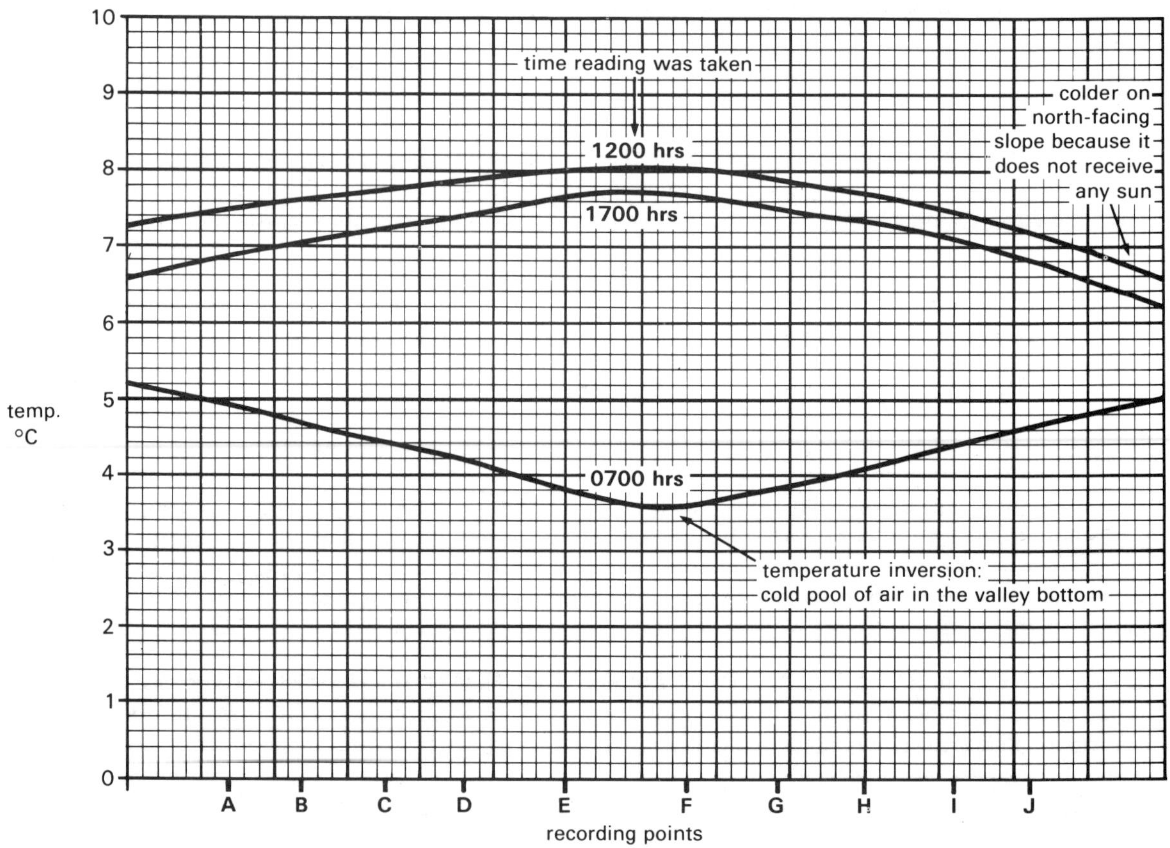

Fig 3.51 (b) Graph to show temperatures recorded on one day, 21st January

Climate is the average condition, over several years, of atmospheric elements such as temperature, humidity, precipitation, air pressure, wind, cloud, and sunshine. The climate of an area varies in detail from place to place; for example it will tend to be warmer and less humid in a built-up area than in a park or field. These differences in local or microclimate can usefully form the basis of a project. Figure 3.51 shows the results of a study of temperature variations across a valley.

Meteorology is the study of weather – the day to day variations in the climatic elements. These variations can also be measured and analysed.

We will first describe the methods by which each of the climatic elements can be measured and then the uses to which such measurements can be put.

Temperature

Temperature may be measured in three ways:

1 Thermometers A 'simple thermometer' is a glass tube filled with alcohol or mercury which expands or contracts with the rise or fall of temperature. Readings may be made in Fahrenheit (°F), but Centigrade (°C) is more usual these days. Readings are usually taken in the shade and away from draughts.

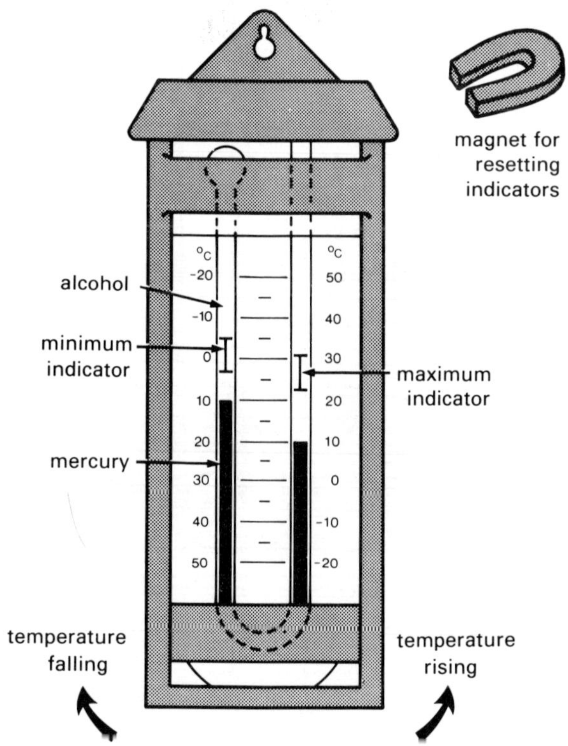

Fig 3.52 Six's maximum and minimum thermometer

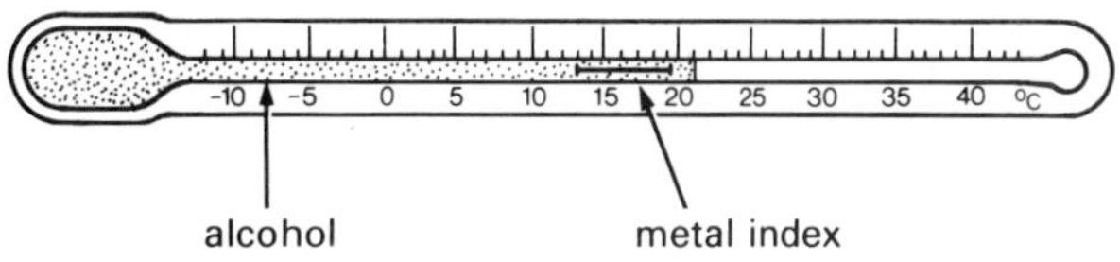

Fig 3.53 Grass minimum thermometer

'Six's maximum and minimum thermometer' (fig. 3.52) records the highest and lowest temperatures in the period for which it is left, normally 24 hours. It is a U-tube with a column of mercury which is pushed round by an expanding column of alcohol when the temperature rises, and pulled back as the alcohol contracts when the temperature falls. Both ends of the mercury column push a small metal indicator before them, and this is left behind when the mercury recedes. Readings are taken at the bottom of each indicator, maximum temperatures on the right-hand column, minimum temperatures on the left. Readings are normally taken once every 24 hours and the indicators are then moved back to rest on the top of the mercury column, using a magnet to stroke them into place.

A 'grass minimum thermometer' is filled with alcohol and has a metal indicator *in* the alcohol (fig. 3.53). When the temperature falls the alcohol contracts and the curved meniscus pulls the indicator with it. When the temperature rises the alcohol flows back past the indicator, leaving it recording the lowest temperature. The thermometer is covered by a glass shield and put on the ground at night. It records the coldest night temperature, which is usually lower than that above the ground.

2 Thermograph In a thermograph (fig. 3.54) the temperature is automatically recorded on a paper strip wrapped around a clockwork revolving drum. It is

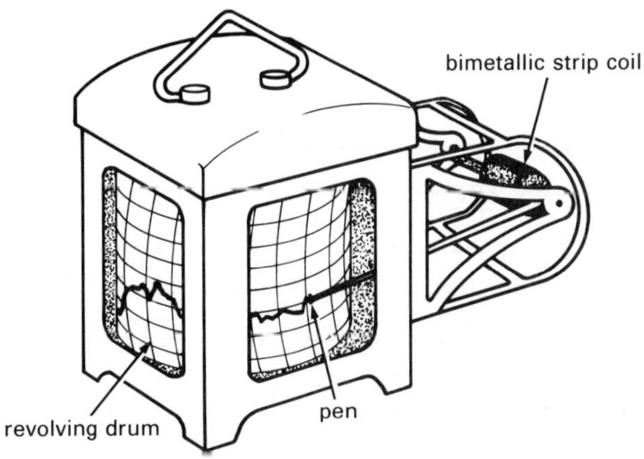

Fig 3.54 Thermograph

thus possible to obtain a continuous record of temperature changes over time. The thermograph is operated by the movement of a bimetallic strip. Two pieces of metal with different coefficients of expansion (i.e. one piece of metal expands at a faster rate than the other) are attached to each other so that when the temperature rises the strip bends (fig. 3.55). This is in turn fixed to an arm on the end of which is a pen which inscribes on the revolving drum.

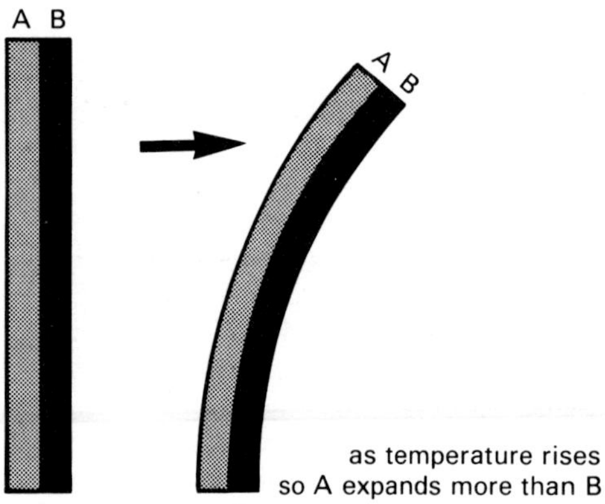

Fig 3.55 The principle of the bimetallic strip

3 Electrical measuring instruments It may be possible for your physics department to provide you with two electrical instruments for measuring temperature, the 'thermistor' and the 'thermocouple'. It is possible to make these yourself but because this often involves a large amount of clumsy apparatus it is better to obtain a professionally made instrument if you can.

A thermistor measures electrical resistance through a capsule containing a substance whose resistance falls as the temperature rises. Thus, the higher the temperature the more electricity passes through. A thermocouple, on the other hand, measures the electromotive force (an electrical current produced between bodies of different temperatures) between two wires made of different metals (such as copper and iron). One wire is kept at freezing point in a thermos of ice, the other reflects the outside temperature. The temperature difference between them generates the electromotive force which is calibrated against a temperature scale.

Both these instruments have a number of advantages, notably that they are very accurate and can be used for measuring the temperature of air, liquids and solids. They are therefore ideal for the study of microclimates where small differences of temperature occur over small areas.

Humidity

Humidity can be measured in three ways:

1 Wet-and-dry-bulb hygrometer (psychrometer) Two ordinary thermometers are mounted on a stand (fig. 3.56). The bulb of one is wrapped in muslin, the end of which is dipped in a small container of water. The muslin around this 'wet bulb' is thus kept damp. The relative humidity of air is the amount of water vapour the air is holding as a percentage of that which it *would* hold if the air was completely saturated. If the relative humidity is less than 100% water evaporates from the muslin around the wet bulb, thereby cooling it. We record the temperatures of the wet and dry bulbs and consult a table or a special slide-rule of values to convert this into the relative humidity.

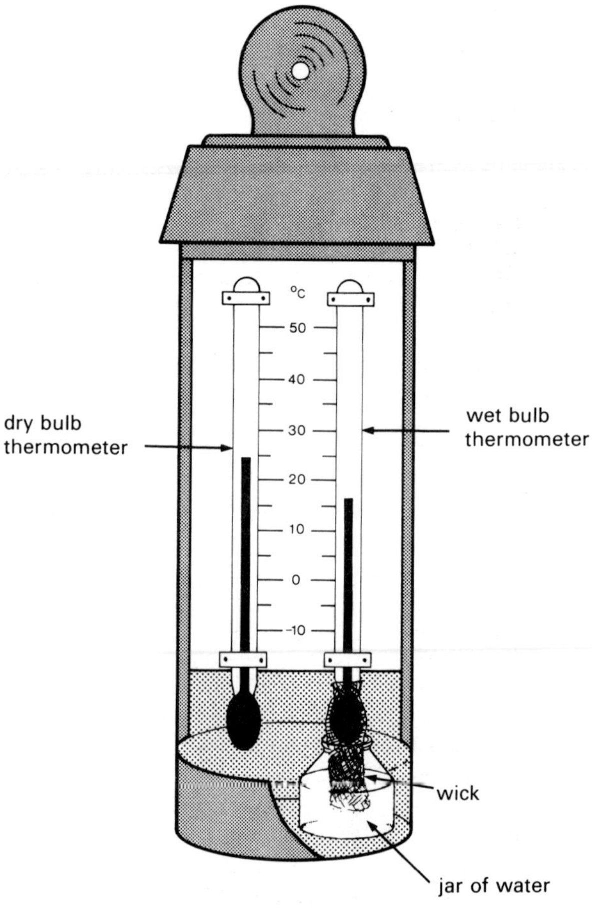

Fig 3.56 Wet-and-dry-bulb hygrometer

2 Whirling psychrometer This comprises wet and dry bulb thermometers mounted on a frame which can be swung round like a football rattle (fig. 3.57). In so swinging air circulates rapidly around the wet bulb, accelerating the rate of evaporation. Readings are taken and converted into relative humidity from tables or a slide-rule.

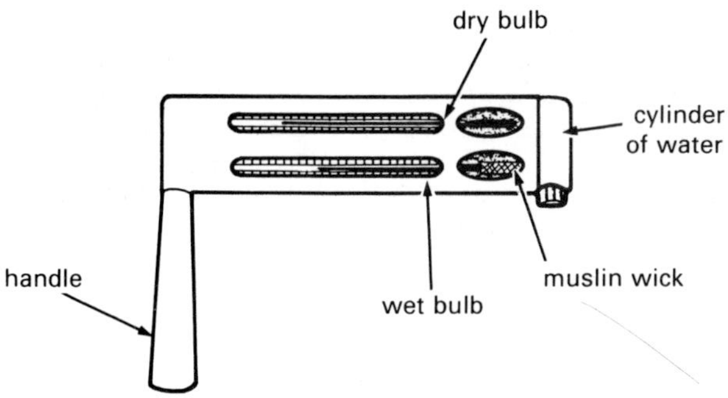

Fig 3.57 Whirling psychrometer

3 Hygrograph A hygrograph is an automatic recording gauge which records humidity on a revolving drum, like the barograph in figure 3.61. This gives a continuous record of humidity levels over time. The hygrograph uses specially treated hair which expands or contracts with varying humidity levels and is connected to an arm and inked pen.

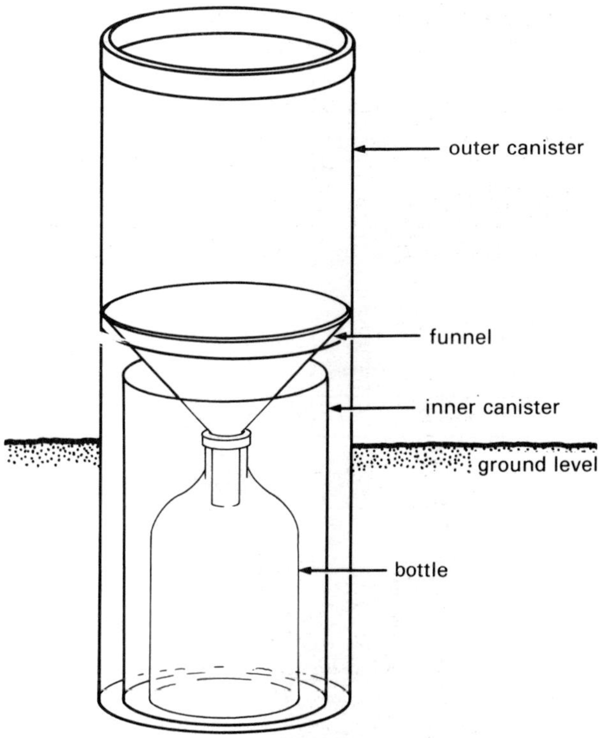

Fig 3.58 Non-recording rain gauge

Precipitation

Precipitation is measured by two types of instrument:

1 Non-recording gauge This is simply a cylinder (canister) which collects rain as it falls (fig. 3.58). If the collecting jar has the same diameter as the cylinder about it the water depth can be measured directly from it. Usually, however, the water is poured into a glass measuring cylinder and the level read off from this. In this case the collecting gauge and the measuring jar must be purchased together as a matching pair.

Snow or hail falling in the gauge must be melted before taking any reading. Readings should be taken *daily* and amounts expressed in millimetres.

2 Recording gauge This automatically records rainfall on a revolving drum. It tells you exactly when the precipitation fell, rainfall intensity, and the duration and frequency of storms. Various methods are used including floats, tipping buckets and weighing collectors. All are expensive.

The exact location and design of the gauge will probably influence the amount of precipitation it catches. It is thus desirable to have *several* gauges if you are calculating precipitation amounts over a broad area, including some under trees.

Air pressure

Air pressure may be measured in three ways:

1 Mercury barometer (Fortin barometer) The mercury barometer (fig. 3.59) comprises a column of mercury in a glass tube. The air above the tube exerts pressure on this mercury and its level varies accordingly. The greater the pressure, the higher the mercury column. The height of the mercury column is read off in millimetres. Reference to a set of tables converts this to millibars (mb), the standard units of pressure. The average pressure in Britain is about 1013 mb, but pressure may vary between about 970 mb and 1040 mb.

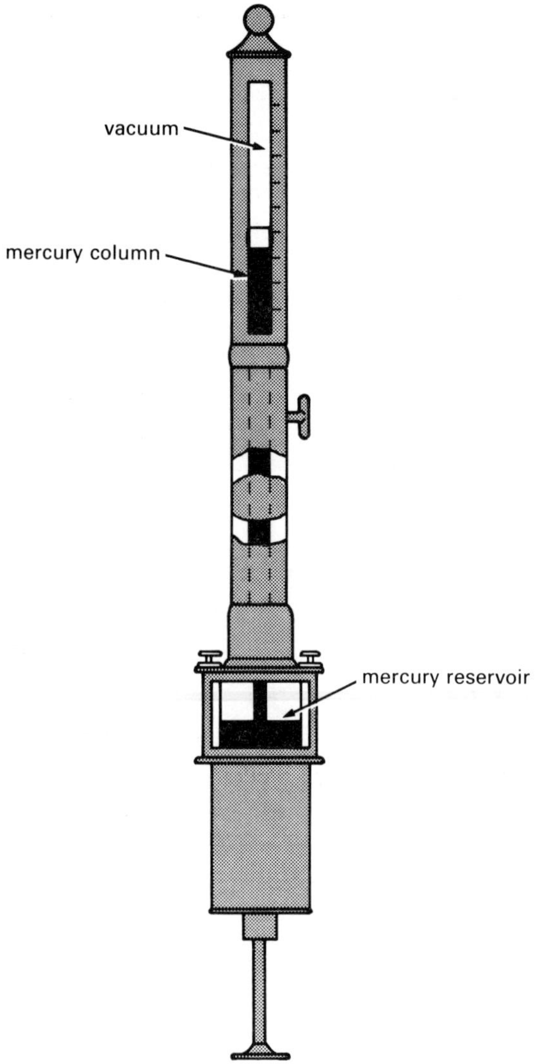

Fig 3.59 Mercury barometer (Fortin type)

2 Aneroid barometer An aneroid barometer (fig. 3.60) is composed of a series of hollow discs within which there is a partial vacuum. As air pressure falls the discs expand. This moves an arm which indicates the pressure on a dial, often using such terms as 'change' or 'fair'.

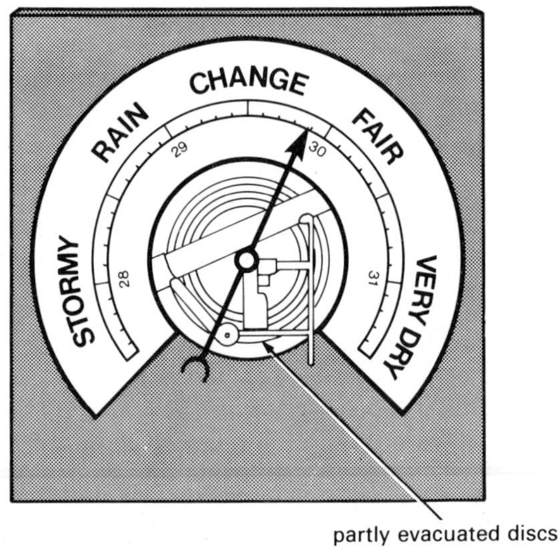

Fig 3.60 Aneroid barometer

3 Barograph A barograph (fig. 3.61) is an aneroid barometer connected to an arm and inked pen which records changes in pressure continuously on a revolving drum. The drum may revolve once a day or, more usually, once a week.

Fig 3.61 Barograph

Wind

Wind speed may be measured in four ways:

1 A revolving cup anemometer This comprises three revolving cups connected to a recording meter which counts the number of rotations in a given time period (fig. 3.62). The meter may then convert this to wind speed in kilometres per hour. Revolving cup anemometers may be fixed or hand-held.

Fig 3.62 Revolving cup anemometer

Beaufort number	Descriptive title	Effect on land features
0	Calm	Smoke rises vertically
1	Light air	Direction shown by smoke but not wind vanes
2	Light breeze	Wind felt on face; leaves rustle; vane moved
3	Gentle breeze	Leaves and twigs in constant motion
4	Moderate breeze	Raises dust and paper; small branches moved
5	Fresh breeze	Small trees begin to sway
6	Strong breeze	Large branches in motion; whistling in telephone wires
7	Moderate gale	Whole trees in motion
8	Fresh gale	Breaks twigs off trees
9	Strong gale	Slight structural damage to roofs, etc.
10	Whole gale	Trees uprooted; considerable structural damage
11	Storm	Widespread damage
12	Hurricane	Widespread devastation

Fig 3.63 The Beaufort Scale of wind force

2 A pressure tube (Dines) anemometer In this case wind passes over or through a tube creating differential pressures as it does so. This operates an arm which moves an inked pen over a revolving drum, giving a continuous record of changes in wind speed. The anemometer is kept pointing into the wind by a weather vane, to which it is attached.

3 A ventimeter is similarly a tube over which the wind passes. The faster the wind, the greater the reduction in pressure in the tube. This causes a pointer to rise up the tube, recording wind speed. Ventimeters are cheap and can be bought in shops selling sailing equipment.

4 By observing the effects of the wind on objects around. This method is based on the Beaufort Scale which is detailed in figure 3.63.

Wind direction is indicated by a wind vane (fig. 3.64) which should be mounted away from obstacles

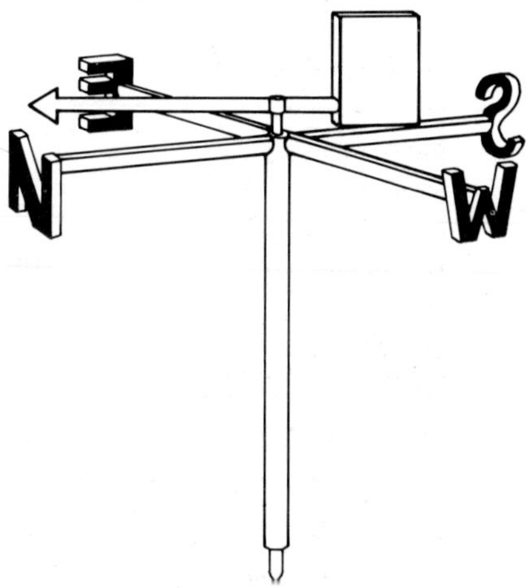

Fig 3.64 Wind vane

likely to interfere with wind flows. The pointer indicates the direction from which the wind is blowing, and wind direction is always quoted as that direction *from which* it blows. An 'easterly wind', therefore, blows *from* the east.

Sometimes wind blows in different directions at different altitudes. This fact can be observed by watching clouds at different levels and noting their direction of drift.

Cloud

Cloud amount is recorded by observing the sky and estimating what proportion of the sky is covered by cloud. This is expressed in *eighths* covered and the units are called 'oktas'. Thus, sky which is half covered by cloud is said to have four oktas cloud coverage. Sometimes the sky cannot be seen because of fog or rain and this fact ('sky obscured') is recorded.

Cloud type is recorded by looking at the clouds and comparing them with a pictorial chart of cloud types (fig. 3.65).

Sunshine

There are several types of sunshine recorders. The most common type uses a glass ball mounted on a frame (fig. 3.66). The sun's rays are focused by the ball onto a strip of special paper behind it. When it

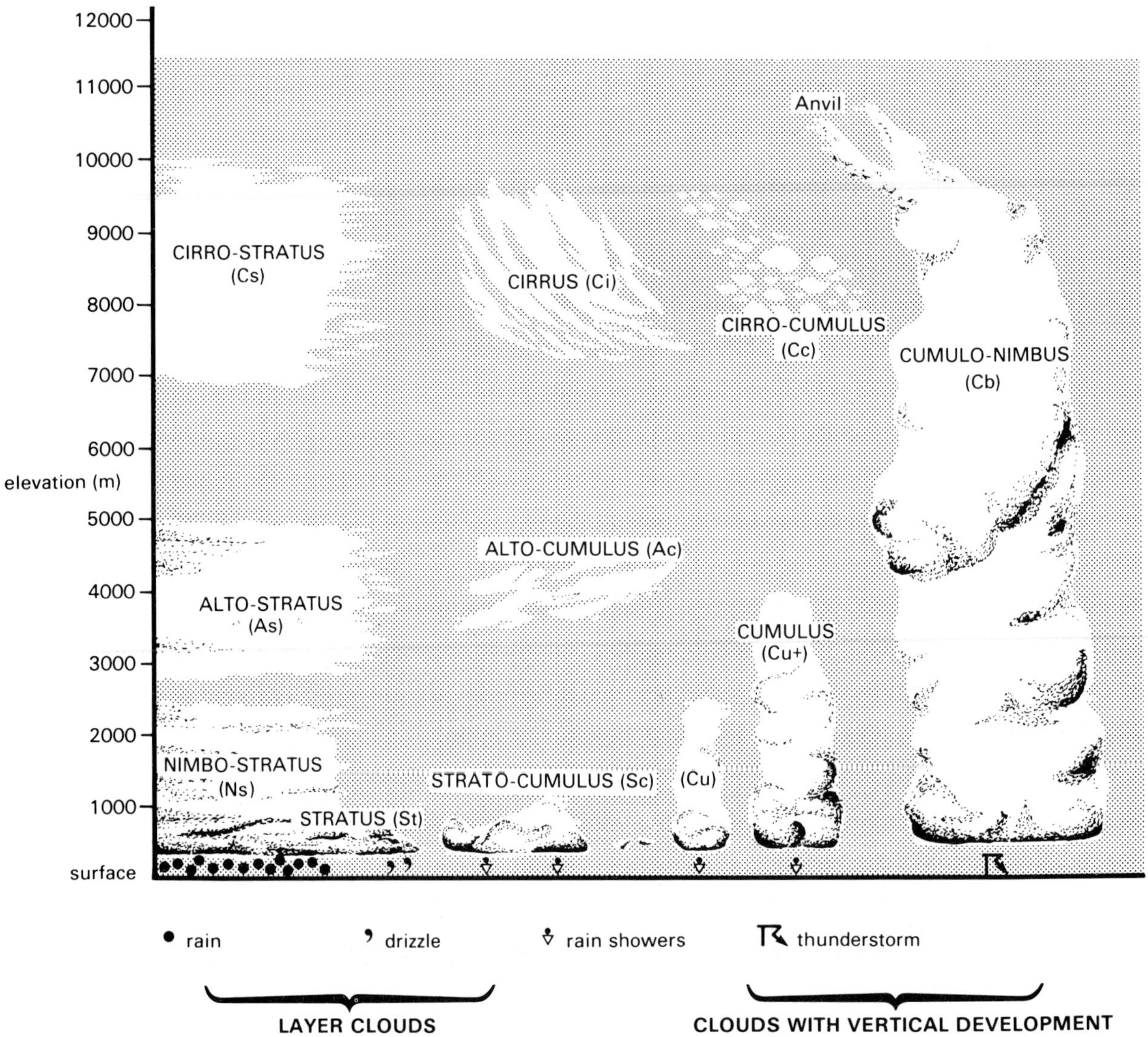

Fig 3.65 Cloud types

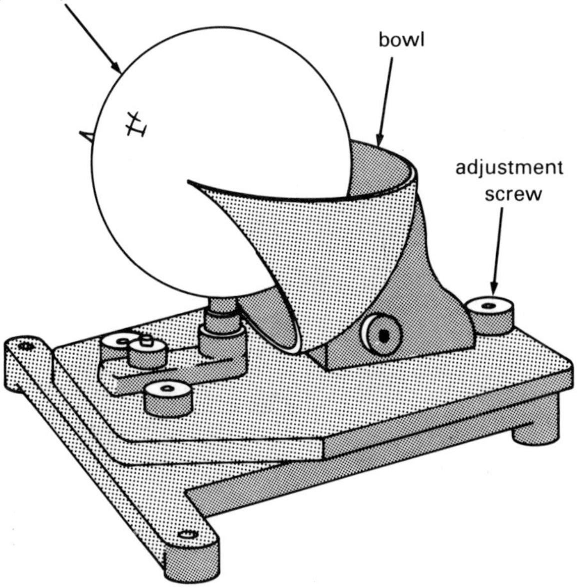

Fig 3.66 Sunshine recorder

shines the sun burns a line along the paper as it moves round in the sky. The total length of burnt line indicates how many hours in the day the sun has shone.

A similar system involves a black tube with a pinhole pricked into it. Sun rays pass through this and mark special paper in the same way.

Meteorological recording

A permanent 'weather station' should be set up in an area away from trees and buildings so that it is not greatly affected by very local conditions. A special Stevenson Screen (fig. 3.67) is used to provide a shaded and sheltered atmosphere in which to keep the thermometers, at a height of 1.2 m above the ground. This standardisation of conditions enables us to compare readings with those of other stations which also use Stevenson Screens. The sunshine recorder may be kept on the roof of the screen, while the rain gauge and grass minimum thermometer will usually be nearby on the ground. Anemometers and wind vanes are ideally mounted on the tops of buildings and barometers kept inside.

Fig 3.67 A Stevenson Screen weather station

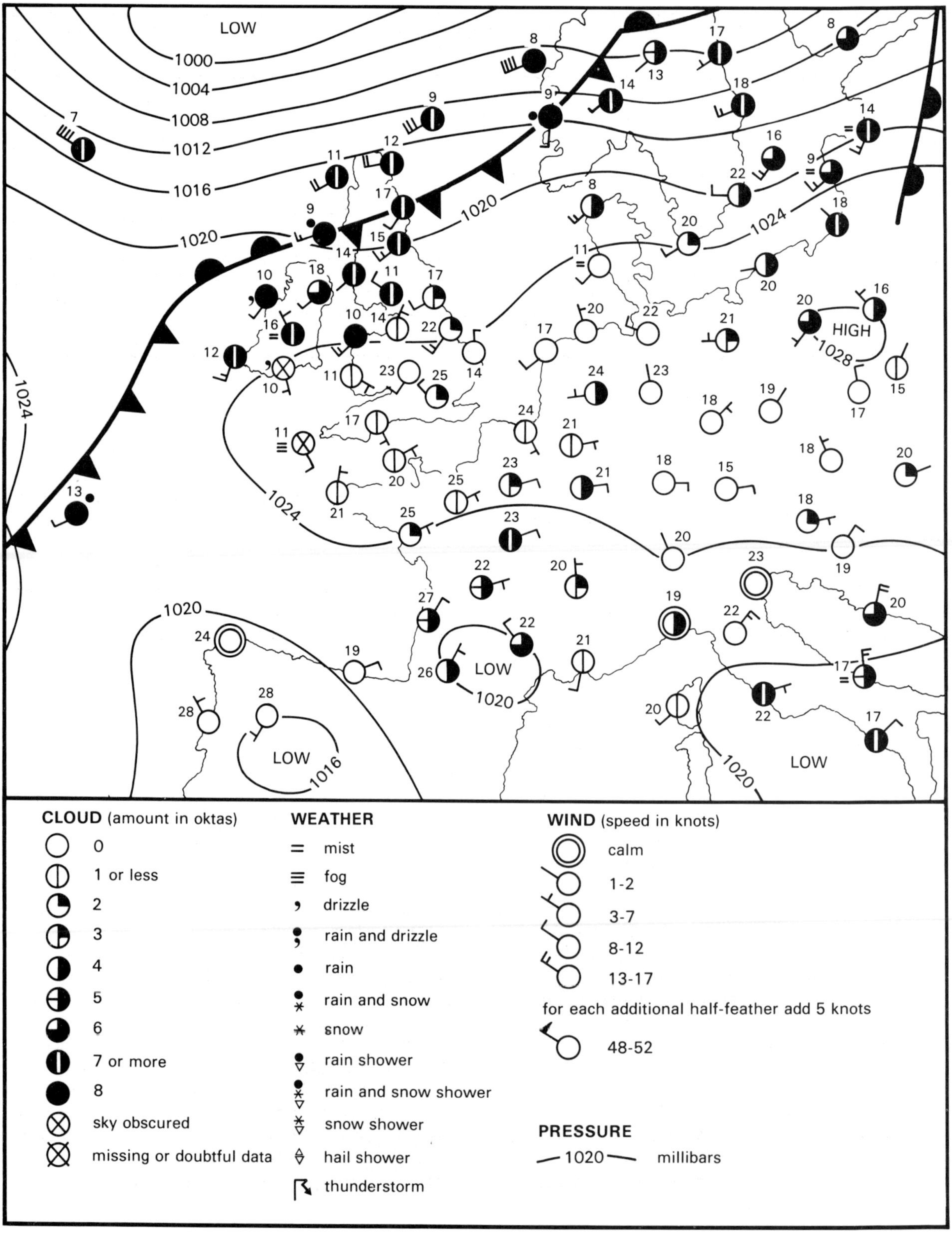

Fig 3.68 A synoptic chart

date	time	temp °C	wet bulb °C	R.H.	max. temp °C	grass min. °C	precipitation (mm)	sunshine (hours)	cloud cover (oktas)	cloud type	wind strength (Beaufort)	wind direction	pressure (mb)	weather
3rd Feb	0900	12	10	78	16	0	0	7	2	Ci	0	—	1022	Fine
4th Feb	0900	10	9.5	94	16	0	0	7	2	As	2	E	1020	Fine
5th Feb	0900	9	8.7	95	13	3	0	4	4	Cu/As	4	S.W	1013	Dull
6th Feb	0900	9	8.7	95	13	2	4	3	6	Cu	5	W	995	Wet
7th Feb	0900	10	9.6	95	14	3	3	3	7	Cu+	5	N	1000	Clearing
8th Feb	0900	12	10.1	79	14	0	0	4	5	As	3	N	1020	Dull
9th Feb	0900	12	10.2	79	14	1	0	5	4	Sc	3	W	1028	Fine

Fig 3.69 Meteorological record sheet

Readings should be taken once or twice a day and noted down on a record sheet (fig. 3.69). The times chosen should be the same every day to allow comparison between days.

Synoptic charts (fig. 3.68) showing the weather pattern over the whole country are most useful for discovering exactly which weather systems (anticyclones or depressions) were responsible for the changes in conditions you recorded.

Synoptic charts and satellite photographs

Synoptic charts are maps summarising the weather for one moment of time. They can be found in many daily national newspapers. They are based on charts produced by the Meteorological Office, which collects weather data from land and sea stations. A synoptic chart is shown in figure 3.68 with the symbols explained underneath.

Satellite photographs of cloud patterns help in the production of synoptic charts as the form and movement of clouds tells meteorologists much about present and future weather conditions. Figure 3.70 shows a typical satellite photograph with an accompanying diagram to explain the significance of the cloud patterns shown.

Project suggestions

Two types of meteorology project can be particularly successful:

1 A permanent weather station approach. Record all the elements of weather at one place over a period of one or two weeks. To do this you will need to set up a Stevenson Screen as described above. Relate your readings to the changing pressure systems, using synoptic charts. If data is available for your area over a long period of time, compare your results with those for the same time of year in previous years.

2 A comparison of meteorological conditions over a small area. Record meteorological data at several places *at the same time* to show how conditions are affected by, for example, vegetation, water bodies, buildings or altitude. Figure 3.71 shows the position of the recording points at different distances from a lake and some of the results obtained. These results clearly show the influence of the water on local climate, depressing temperatures and increasing humidity. You could similarly compare conditions:

- in a town and beyond the town.
- within your school or college grounds.
- at ground level with those at increasing heights above the ground.
- in an open field with those in a wood.
- across a river valley.

The number of places you record at will normally depend on the number of instruments you can obtain. It is important to try and take the readings at approximately the same time so that they can be compared fairly. Note the influence of the overall weather conditions on your results.

Readings can be taken twice a day for several days but it is probably better to take them once an hour for only one or two days. In this way you are comparing the microclimates of different places and of different times of the day. Tests of significance (page 98) should be carried out on your data to see whether any difference between these points is truly significant or whether it could have occurred by chance alone.

Fig 3.70 (a) Satellite photograph of a depression over NW Europe

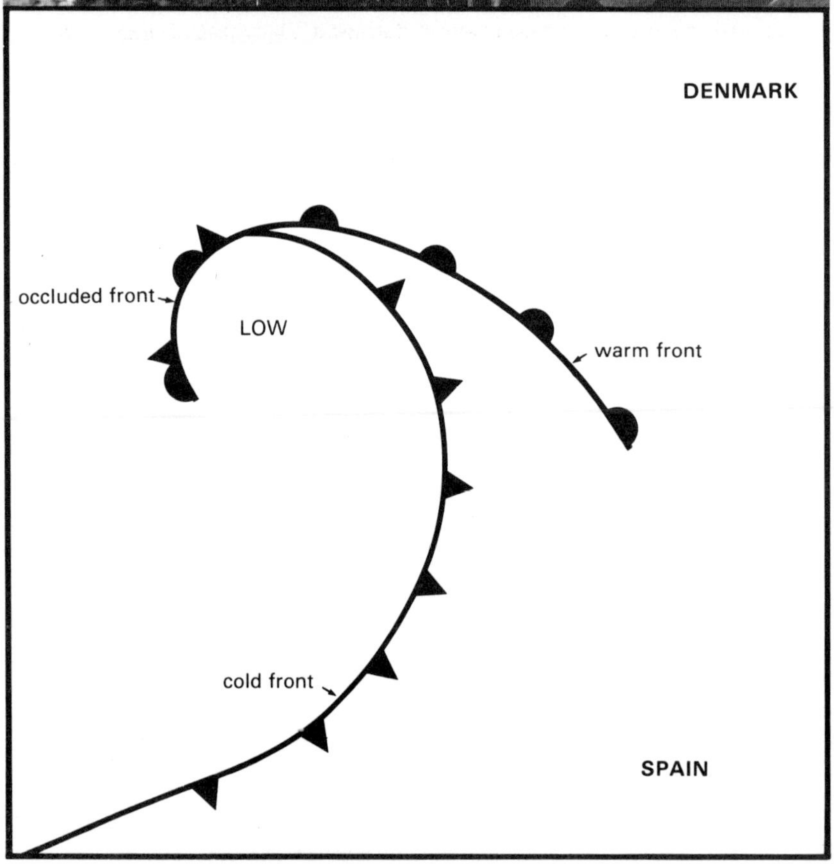

Fig 3.70 (b) Interpretation of the satellite photograph

	STATION A	B	C
Max. temperature (°C)			
DAY 1	21.0	19.6	18.8
2	23.2	22.6	21.8
3	23.4	22.6	22.0
4	23.8	22.2	21.6
Min. temperature(°C)			
DAY 1	8.6	8.0	8.0
2	13.0	11.6	11.5
3	11.4	11.0	10.8
4	10.2	10.0	9.8
Relative humidity (%)			
DAY 1	76.9	77.3	78.2
2	74.3	77.4	78.3
3	75.2	77.6	79.4
4	75.1	78.1	79.4

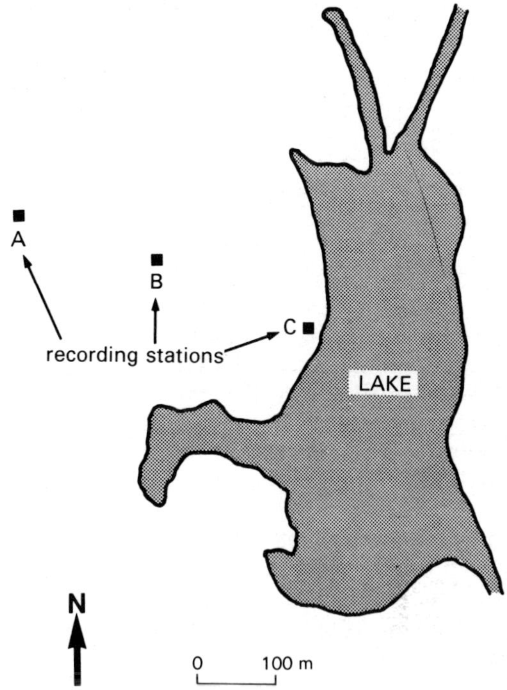

Fig 3.71 Three recording stations at different distances away from a lake
Temperature readings were taken from the stations at 8.30 a.m. on four consecutive days.
Maximum and minimum readings refer to temperatures experienced during the previous 24 hours.

4 Human geography

Although human activities are often hard to measure, a number of techniques and sources of data, which make some sense out of the seeming confusion of man's behaviour, are now available. In this chapter section 4.1 gives details of one of the techniques available – that of carrying out questionnaires and interviews. Sections 4.2 to 4.6 each take a particular topic and outline details of the appropriate data sources, followed by some suggestions for projects. Finally, section 4.7 briefly covers the general catalogues of data sources which may be of use to you.

Before you begin to follow up any of the data sources outlined in this chapter it is best to be aware of the limitations of published data. Of particular importance is the problem of the *areal units*. All data is collected and published for given areas such as counties, parishes or units specially devised for the purpose. Sometimes these units are too large and conceal huge internal variations. Sometimes they have changed over time so that data for one period cannot be fairly compared with that from another because the areal units are different. Often we want to compare data which has been collected from within two different sets of areal units, and this is also a problem. Finally, the boundaries of the units may be badly drawn. Data for towns, for example, may be published for areal units which were fixed some time ago. If the town has since expanded beyond these boundaries the data will fail to include information about the outer suburbs.

4.1 Questionnaires and interviews

Questionnaires are the most important primary data source available in human geography. Two examples of a questionnaire are given in figure 4.1: one is good and the other bad. They illustrate some of the most important points to remember when using questionnaires:

1 Try to decide how many questionnaires you are going to do before starting. Do not try to do too many nor so few that the results you obtain *could* have occurred by chance.

2 Plan carefully *where* and *when* you are going to conduct the questionnaire. If, for example, you are hoping to interview people about their shopping habits, you must decide whether you wish to do a house-to-house survey or stop people in the High Street. You will have to decide which days of the week to do it on and which times of day. Where you stand along the High Street will also matter. All these things affect the results and you *must* explain your choice when you write up your project.

3 People do not like answering questionnaires. They must be short and never include offensive questions concerning age or income. People are also reluctant to give their address so ask for 'the general area where you live'. Some people will refuse to answer your questionnaire. The result will inevitably be biased towards those people who are willing to do it.

4 Decide what *sort* of questionnaire you are going to do. There are various types:
   a Standing in the street and catching passers-by.
   b House-to-house surveys, which are biased against working people if conducted during the day.
   c Questionnaires sent to people and picked up from them at a later date.
   d Questionnaires sent to people including a stamped addressed envelope in the hope (usually vain) that people will send it back to you.

   You may wish to interview specific individuals, such as the owners of factories or shopkeepers. Such interviews should always be arranged in advance by letter or telephone.

5 The format of questions in questionnaires should be carefully planned so that the respondent is put at ease at the beginning of the interview. More probing questions should be left to the end.

6 Always conduct a pilot study in which the questionnaire is given to a small sample. Such a trial run enables inconsistencies to be noted and possible sources of misunderstanding and confusion to be eliminated in subsequent versions.

7 Always dress smartly and be polite. Do not, however, expect people to be polite to you – they are suspicious of all such surveys. Begin by explaining that you are a student doing a geography project.

8 Do not ask questions which 'expect' a certain answer, such as question 8 in the bad questionnaire (fig. 4.1).

9 Above all, know *why* you are asking every question you use and do not *omit* anything you must know. You cannot go back later, find the same people and fill in any gaps.

A

Good Questionnaire

Introduction: 'Excuse me, I am doing a school geography project. Could I ask you one or two quick questions about where you go shopping?'

1 How often do you come shopping in this town centre?
More than once a week ☐ Weekly ☐ Occasionally ☐

2 How do you travel here?
Walk ☐ Car ☐ Bus ☐ Train/Tube ☐
Other ....

3 Roughly where do you live? ....

4 Why do you come here rather than any other shopping centre?
Near to home ☐
Near to work ☐
More choice ☐
Pleasant environment ☐
Other ....

5 What sort of things do you normally buy here?
Groceries ☐ Clothes/shoes ☐ Everything ☐
Other ....

6 Do you ever shop anywhere else, and if so where?....

7 Why do you go shopping there? ....

8 What do you buy there?..........

9 Sex: M ☐ F ☐ Age (estimate): under 20 ☐ 20–30 ☐ 30–60 ☐ over 60 ☐

'Thank you very much for your help.'

B

Bad Questionnaire

Introduction: 'Excuse me, but I wonder if I could ask you some questions?'

1 Where do you live?..........

2 How do you get here? ....

3 Do you come shopping here often? ....

4 Why do you come here?..........

5 Do you buy high- or low-order goods here?..........

6 Is this a good shopping centre and if so, why? ....

7 Where else do you go shopping?..........

8 Do you shop there because it is cheaper or nearer to your home?..........

9 How old are you?..........

'Right, that's it then.'

Fig 4.1 Two questionnaires, one good and one bad. Both were designed to find out about the shopping habits of people and were administered in a town centre.

Questionnaires and interviews are valuable because they provide up to date information about individuals which is not normally available from published sources. They are particularly important when it is necessary to find out people's opinions and motives, for example *why* someone moved house or *how* a decision was made to open a shop in a town.

4.2 Population and settlements

Data sources

1 The National Census The National Census has been held in Britain every ten years since 1801, except 1941. The last census was in 1981. It is published in tables which can be read at the Office of Population Censuses and Surveys (10 Kingsway, London WC2), in libraries and local authority planning departments.

Three main kinds of statistical table are produced: County Tables for districts and counties, Small Area Statistics for areas smaller than local authority districts, and subject volumes which give national and regional figures on special subjects such as travel to work or birthplace. It is not possible to obtain census information about very small areas, such as individual streets.

The actual enumerators' returns can be seen after 100 years, and these give details for *individuals* rather than areas. It is thus possible to obtain highly detailed records from the 1881 census and those preceding it. These can be read at the Public Record Office, Chancery Lane, London WC2, and some local libraries may have copies.

The main topics covered by the 1981 census are as follows:

Sex composition	Occupation
Age structure	Employment status
Marital status	Educational qualifications
Fertility	Location of work
Household structure	Method of travelling to work
Population numbers	Housing tenure
Country of birth	Housing amenities and space
Population movement	Ownership of motor vehicles

The data (in the form of Small Area Statistics) is published at three scales: boroughs, wards and enumeration districts. Boroughs are the largest with a population size ranging from 60 000 to 200 000. Wards range from 8000 to 30 000 and enumeration districts from 200 to 1000 people. It is essential to obtain a base-map of these areas before beginning to interpret the data given for them. Not all data is available at every scale; indeed some figures are only published for large areas such as counties or towns with populations over 50 000.

2 Migration Tables and Reports In addition to the Census, Migration Tables can be purchased from Her Majesty's Stationery Office (HMSO) at 49 High Holborn, WC1V 6HB. These are based on a 10% sample of all households and give, for each county, county borough and town over 50 000, information on migration in the five years prior to the Census. This includes details of the area of former residence of migrants, the type of move, their sex, age, marital status, socio-economic status, household tenure, housing amenities and population density.

Migration Regional Reports (HMSO) cover similar data for Economic Planning Regions in England and Wales. These regions are large: for example the Northern Economic Planning Region covers the whole of the northern part of England from Cumbria in the west to Newcastle in the east.

3 Other Central Government population statistics The Office of Population Censuses and Surveys (OPCS) also publishes, through HMSO, up to date statistics on a variety of topics such as birth rates and mortality. These include 'Population Trends' (a quarterly publication), the 'OPCS Reference Series' and 'OPCS Monitors'. The Central Statistical Office's periodical 'Social Trends' (HMSO) is also widely used for this type of data.

4 Parish Registers Parish Registers, recording weddings, baptisms and burials in local Anglican churches, have been kept in Britain since 1538. They are one of the most readily accessible data sources for an historical analysis of population. The information given in the registers is as follows:

Baptisms
Date of birth
Date of baptism
Name of child and parents
Address
Father's occupation

Marriages (fig. 4.2)
Date of marriage
Name of marrying couple
Age of marrying couple
Occupations of marrying couple
Address of couple before marriage
Name of fathers
Occupation of fathers

Burials
Date of burial
Name of deceased
Age at death
Address at death

Fig 4.2 A sample page from a Marriage Register

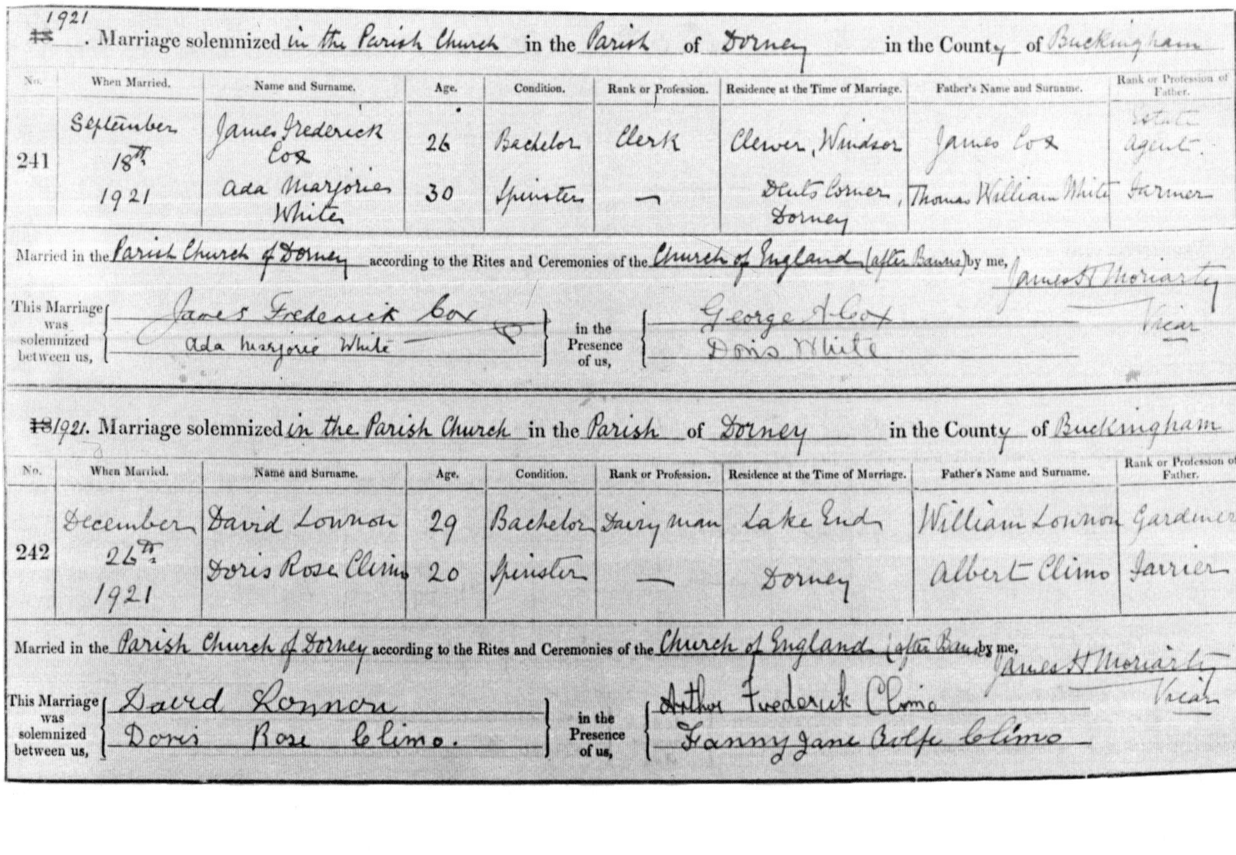

1921. Marriage solemnized in the Parish Church in the Parish of Dorney in the County of Buckingham

No.	When Married.	Name and Surname.	Age.	Condition.	Rank or Profession.	Residence at the Time of Marriage.	Father's Name and Surname.	Rank or Profession of Father.
241	September 18th 1921	James Frederick Cox	26	Bachelor	Clerk	Clewer, Windsor	James Cox	Estate Agent
		Ada Marjorie White	30	Spinster	—	Dents Corner, Dorney	Thomas William White	Farmer

Married in the Parish Church of Dorney according to the Rites and Ceremonies of the Church of England (after Banns) by me, James H Moriarty Vicar

This Marriage was solemnized between us, James Frederick Cox, Ada Marjorie White, in the Presence of us, George A Cox, Doris White

1921. Marriage solemnized in the Parish Church in the Parish of Dorney in the County of Buckingham

No.	When Married.	Name and Surname.	Age.	Condition.	Rank or Profession.	Residence at the Time of Marriage.	Father's Name and Surname.	Rank or Profession of Father.
242	December 26th 1921	David Lonnon	29	Bachelor	Dairy man	Lake End	William Lonnon	Gardener
		Doris Rose Climo	20	Spinster	—	Dorney	Albert Climo	Farrier

Married in the Parish Church of Dorney according to the Rites and Ceremonies of the Church of England (after Banns) by me, James H Moriarty Vicar

This Marriage was solemnized between us, David Lonnon, Doris Rose Climo, in the Presence of us, Arthur Frederick Climo, Fanny Jane Rolfe Climo

From this it is possible to study changes over time of average age of marriage, birth rates, death rates, the distance between the pre-marital addresses of partners, the structure of occupations in a parish and life expectancy. Parish Registers are held by the local Anglican clergyman, who will normally allow you to see them, or by the County Record Office.

The main problems with Parish Registers as a data source are as follows:

a It is not possible to tell the total parish population from them. This figure must be obtained from another source if birth and death rates are to be calculated.

b Gaps often appear in the records and many registers have been lost altogether.

c Before this century the registers did not include dissenters from the Anglican Church, those who could not or would not pay the necessary fees, and, in the case of burials, they omitted suicides, criminals and the unbaptised. In the present century many people do not get married in church, preferring a Registry Office, many children are not baptised and not everybody receives a Christian burial. Missing data may be obtained from Registry Offices and the County Record Office.

d The areas of parishes may change over time.

5 Kelly's Directories These were published annually after 1845, for counties (divided by parishes) up to 1946 and for larger towns up to 1976. They include lists of residents in a settlement, recorded in order down each side of their street. Comparing directories over a number of years can thus help studies of migration and housing turnover.

6 Electoral Registers Electoral Registers (fig. 4.3) are compiled every year by local authorities. They list the name and address, by street, of all people eligible to vote. By comparing registers for two years it is possible to see how many people have moved house during the intervening period. Following this up with a questionnaire designed to find out where people have moved from and why provides the basis for a project on local migration. Electoral Registers can be seen in local authority offices and main libraries.

7 Rateable values and rate books Every building has a rateable value, which represents a local authority officer's assessment of what would be a fair annual rent for that building. It is based on the condition of the building, its location and the value of the land it is on.

WREXHAM WEST CONSTITUENCY
POLLING DISTRICT

NUMBER	NAME AND ADDRESS
	001-ASH ROAD
1	WALTERS GWEN
2	WALTERS WILLIAM N.
3	JOHNSON ANNE M.
4	JOHNSON MALCOLM E.
5	ANSELL ELSIE W.
6	ANSELL JACK
7	ANSELL JOHN
8	SMEWING VIOLET E.
9	STOKES ALISON
10	STOKES PETER
11	MILNE DONALD
12	PALMER IRENE
13	PALMER MERVYN H.
14	EVANS GWILYM T.
15	EVANS JOAN A.
16	THOMAS DORIS G.
17	THOMAS ROBERT J.
18	BRAYFORD MARTIN C.
19	BRAYFORD SUZETTE H.
20	COOPER ALISON E.
21	COOPER BARBARA P.
22	COOPER EDWARD A.
23	BARLOWSKI ELIZABETH
24	BARLOWSKI FRANCISZEK
25	BARLOWSKI MARIA
26	BARLOWSKI RICHARD F.
27	FINDLAY BRIDGET C.
28	FINDLAY STEPHEN J.
29	JUDGE VALERIE M.
30	SMITH REGINALD I.
31	SMITH THELMA
32	WALSH DENNIS J.
33	WALSH MARIE
34	O'ROURKE CHRISTOPHER C.
35	O'ROURKE GEOFFREY C.T.
36	O'ROURKE PATRICIA M.
37	O'ROURKE SALLY E.
38	STEIN MICHAEL J.
39	STEIN PATRICIA E.
40	PEET ANTHONY W.
41	PEET CHRISTINE J

Fig 4.3 Extract from an Electoral Register

Rateable values can be obtained from local authority rates offices, and every member of the public has a right to inspect them. Figure 4.4 is an example. The 'Gross Value' is the full rateable value. The 'Rateable Value' column on the right is used to calculate the rates the property holder must pay. The difference between the two represents the likely annual cost of maintaining the property.

Rateable values can be used for showing the distribution of property values in a town and thereby analysing land uses. For an historical analysis rate books can often be obtained for periods going back

TRURO B.C. RATING DISTRICT					
ANALYSIS CODE	DESCRIPTION	ADDRESS (& NAME OF OCCUPIER, OR OWNER 'O' IF REQUIRED FOR IDENTIFICATION)	GROSS VALUE	RATEABLE VALUE	REFERENCE NO. OF AMENDMENT
		HERITAGE CLOSE	£	£	
1050	Flat	1	330	250	
"	do	2	330	250	
"	do	3	330	250	
"	do	4	330	250	
"	do	5	330	250	
"	do	6	356	271	
"	do	7	330	250	
"	do	8	356	271	
"	do	9	330	250	

Fig 4.4 Extract from a Rateable Values book

at least as far as the early nineteenth century. These say who owned each property, who occupied it, what type of property it was and its rateable value. They can be used to study property ownership and migration patterns.

8 Maps

a Ordnance Survey maps For most fieldwork a necessary preliminary step is to obtain Ordnance Survey maps of the area. These can be used for finding your way around, as a sampling frame on which sample points, lines or areas will be selected, or as a base-map on which other data may be plotted. They are also a valuable guide to land uses, such as the distribution of woodland, marsh and heath. Road patterns in towns often help identify types of housing area before observation is made on the ground. For example, a grid of straight roads with narrow streets often indicates an area of Victorian working class housing, while crescent patterns and cul-de-sacs are more typical of post-war housing.

Ordnance Survey maps also enable one to study the overall pattern of settlements; these can be further analysed using the various methods available for summarising point patterns, discussed in chapter 7.

The following types of map are available:

1:1250 for larger towns and conurbations.

1:2500 for much of Britain except remoter rural areas and larger towns.

1:10 000 and 6 inches to 1 mile for all of Britain except remote parts of Scotland.

1:25 000 (2½ inches to 1 mile) for all except remote parts of Scotland. There are also special tourist 'Outdoor Leisure Maps' at this scale.

1:50 000 (replacing the old 1 inch to 1 mile maps) which cover the whole of Britain. There are also some tourist maps available at this scale.

1:250 000 (4 miles to 1 inch) covering the whole of Britain.

1:625 000 (10 miles to 1 inch) which show items of special interest such as coal and iron, rainfall, vegetation and types of farming.

David and Charles Ltd publish facsimiles of the 1 inch to 1 mile 1801 (first series) maps for England and Wales. Earlier maps are usually kept in the Local Record Office and the reference section of larger local libraries.

b Goad maps Goad maps (published by Charles E. Goad Ltd of 18a Salisbury Square, Old Hatfield, Hertfordshire, AL9 5BE) show the shopping centres of towns and cities with populations over 50 000 in Britain. They also publish shopping centre plans for some suburban shopping centres in large cities. An example is shown in figure 4.5 but this is reduced from the actual scale of 88 feet per inch (1:1056).

Fig 4.5 Extract from a Goad map of Chippenham, Wiltshire

Goad maps show the location of shops with the name and type of shop marked on. This is useful for studies such as an analysis of the distribution of certain shop types or the structure of the Central Business District of a town.

The main drawbacks are threefold:

1 They only show the ground floor shops.
2 They quickly go out of date (although this could be used effectively to show the changing shop structure of a centre).
3 They do not always cover the whole area of shops in the centre of a town, especially if there are small clusters of shops separate from the main body.

c Local authority maps Local authorities may well hold other maps used for some specific purpose, such as those showing plans for future expansion of a village or town centre. It may be possible to purchase copies from the local authority planning office.

d Town street plans Street plans can be bought for most towns and cities. They are useful for finding one's way around and may provide a base-map on which other features, such as house type, can be marked.

9 Local newspapers Newspapers can be used as an information source for population and settlement studies in a number of ways:

a They provide information about the locations of local events and local businesses (through advertisements), giving an indication of the sphere of influence (page 62) of the town concerned.
b Most local newspapers have a large section

devoted to housing for sale. By noting the average selling price and type of property advertised for different areas of the town or city you can build up a picture of the private housing element of the local housing stock.

c All local newspapers provide a coverage of the important planning decisions reached by a local authority and sometimes an analysis of the kinds of conflicts involved in such decisions: future road building, housing and recreation schemes, for example. These may suggest possible projects to you and provide a wealth of information.

Public libraries often keep back-copies of local newspapers.

10 Local authority data Local authorities are a most important data source. Their more useful information includes the following:

a **Planning documents** All local authorities have now produced three types of plan for land uses in their area: 'development plans' of proposed developments, 'structure plans' which deal with outline planning strategy and more detailed 'local plans'.

b **Transportation studies** These provide a wealth of material on car ownership, land use arrangements and traffic flows.

c **Rateable values** (above).

d **Information held by separate departments** on housing, water supply, education, sewerage, population change, etc.

e **Local housing statistics** are published every three months for each local authority. They give the number of dwellings under construction for each of the different tenure categories. They also give information on the number of dwellings demolished and the number of house improvement grants approved.

f **Material published by New Town Development Corporations** Many New Towns in Britain are still run by Development Corporations, based in the town concerned. These corporations were set up at the start of each New Town to coordinate development until the town had reached a satisfactory stage of growth. They invariably publish detailed plans, progress reports and information documents.

Project suggestions

The number of possible projects related to population and settlements is vast. Here we mention just a few, touching on some of the more obvious and easily examined themes.

1 The testing of the relationship between the number of services found in settlements and the populations of those settlements could form the basis of a successful project. To do this you would need to list all the services you are likely to find in a town (fig. 4.6) and record the number of each type of service in each of a variety of different sized settlements. Details of services could be obtained from trade directories or from your own fieldwork. You can obtain the population size of each settlement from census data or from local authority sources. Draw graphs and calculate the Spearman's rank correlation coefficient (page 100) for the relationships you discover. Discuss the *reasons* for these results.

Some settlements, such as dormitory towns for commuters, may have a large population but relatively few services. Others, like coastal resorts, may have a small permanent population and many services. In general, however, the larger the settlement the more services it offers. This hypothesis can be tested.

Supermarket	Jeweller
Small grocer	Record shop
Greengrocer	Sports shop
Butcher	Pet shop
Baker	Art shop
Dairy shop	Antique shop
Fishmonger	Toy shop
Off-licence	Cinema
Confectioner/Newsagent	Bingo hall
Delicatessen	Cafe
Clothing shop	Public house
Haberdashery	Restaurant
Furniture shop	Hotel
Carpet shop	Club
China shop	Laundrette
Hardware shop	Hairdresser
Electrical goods	Undertaker
Chemist	Estate Agent
Cycle and motor accessories	Building Society
Second-hand shop	Solicitor
Gift shop	Betting shop
Camera shop	Travel Agent
Bookshop	Optician
Stationer	Photographer
Florist	Job Centre

Fig 4.6 Services found in a town

2 A study of the Central Business (shopping and office) District of a town is another popular theme. The Central Business District (CBD) of most towns has a number of clear 'gradients' within it: the most expensive shops, the tallest buildings and the greatest number of customers are probably found near the centre. As one moves towards the edge of this CBD the shops become less expensive, building heights fall and the density of shoppers drops. These gradients can be studied in a town of your choice.

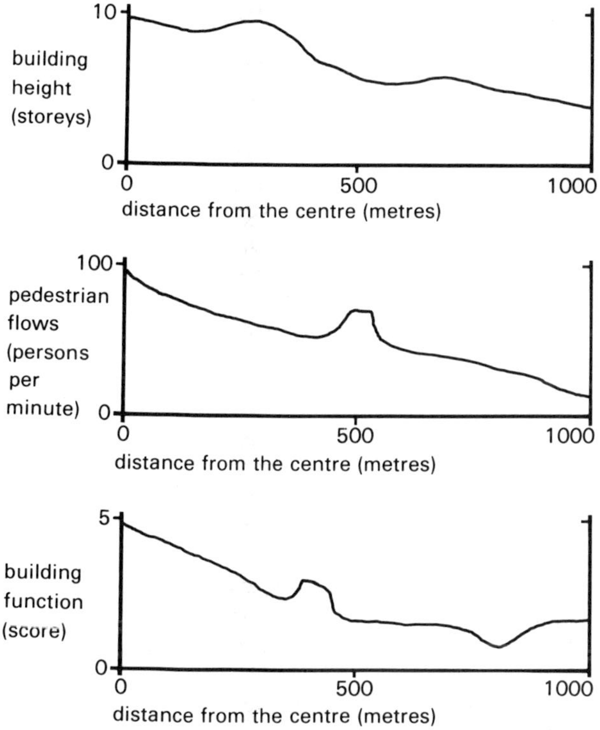

Fig 4.7 Gradients leading out of the Central Business District of Buxton, Derbyshire

Either record data for the whole area of the CBD or collect data for selected roads leading out from the centre to the edge of the CBD. Record the following information: building height (number of storeys), shop floor area or shop frontage (obtained from a Goad map or paced out on the ground), building rateable value (obtainable from the local authority), building function and pedestrian flows at varying distances from the centre. You will probably need to devise a scoring system for the building function, according to how 'important' it is. For example, a large shop or office could be given a score of 5; an ordinary house 1. You will need to take account of 'mixed function' buildings (e.g. a shop with a flat above). Also note any evidence of changing building function over time (fig. 4.8).

All this information can be summarised in graphs (fig. 4.7) to give a series of gradients leading out from the CBD. Spearman's rank correlation coefficient can be calculated for each of the variables, correlated with distance from the centre. Compare the gradients and correlations obtained and see if it is possible to identify a clear *edge* to the CBD or not. Give reasons for the gradients you observe.

Fig 4.8 Changing land uses around the Central Business District: houses changing into shops and offices

3 Another interesting topic is the examination of land use structure in a medium-sized town. All towns exhibit patterns of land use (houses, shops, offices, industry, parks, waste land) which are the result of social, economic, physical and political influences at work over time. Geographers have attempted to summarise these in models (Burgess, Hoyt, Harris and Ullman, Mann). To examine these patterns for yourself you will need to draw up a classification of land uses. For example:

High-status residence	Offices
Middle-status residence	Industry
Low-status residence	Parks and playing fields
Public buildings	Transport facilities
Shops	Waste land

Record these land uses on a street map of your town, obtaining your information from Ordnance Survey maps, land use maps, the Census and from your own fieldwork. Compare your results with the traditional urban models. Attempt to explain *how* your land use patterns were formed.

4 You might like to examine people's perception of residential neighbourhoods (their mental map of the area). Because similar houses tend to be built together in neighbourhoods or estates people feel they 'belong' to a certain part of a town. Ask each person you interview to draw a map of the town, marking on and naming all the residential neighbourhoods they can think of. Ask them if their own neighbourhood has got a name and where they think its boundaries lie. Examine all these replies to establish how clearly people perceive the neighbourhood in which they live. Why are some neighbourhoods more clearly perceived than others? And why are some boundaries a source of greater agreement than others?

5 Another popular topic is the determination of the sphere of influence of a town or shopping centre. Towns and shopping centres do not exist in a vacuum but exert a considerable influence on the region around them beyond the built-up area. People from this region look to the town for work, shops, entertainment and other services; the town looks to the region for a work-force, customers and recreational space. Spheres of influence vary in size and shape, and small spheres 'nest' inside those of larger settlements. It is important to remember, however, that spheres of influence are not a physical entity and cannot therefore be absolutely determined on the ground.

Two approaches lend themselves to this project. You could either use a simple breaking point method or a field survey method. The former assumes that the sphere of influence surrounding each of a pair of competing centres varies in proportion to the size of the centre. The 'breaking point' where these two spheres meet can therefore be found by dividing the distance between the centres in proportion to the size of their populations (fig. 4.9). If breaking points are calculated between the town under study and all of its neighbours the sphere of influence can be delimited by joining up all these points.

The second method involves straightforward fieldwork. Identify functional areas such as school, shop, hospital or library catchment areas by asking the relevant authorities where the people they serve live. If they are unwilling to give details you might try asking people leaving the shop, for example,

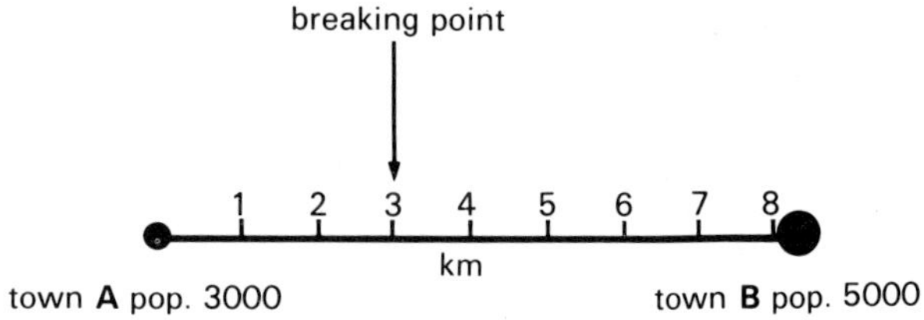

Fig 4.9 Determination of the 'breaking point' between two towns

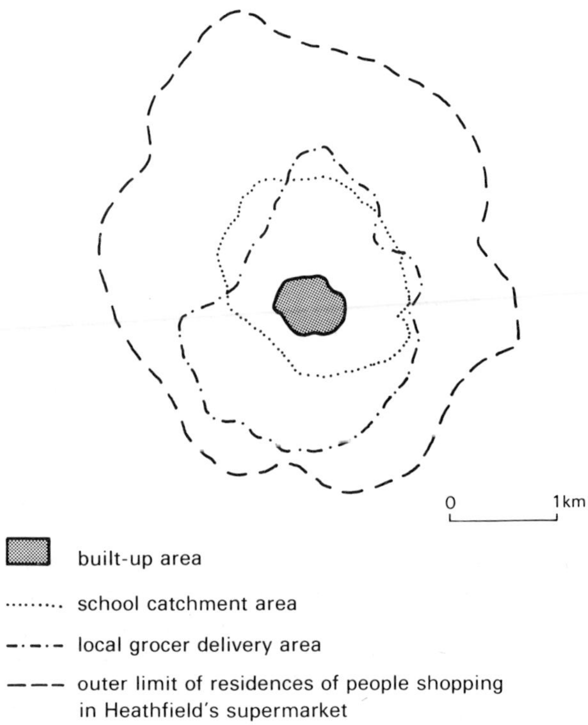

Fig 4.10 The spheres of influence of Stanhope, Co. Durham

approximately where they live, and building up a pattern from their replies. Other possibilities for investigating functional areas are such things as the sources of news in a local newspaper, the location of advertisers in the same, the route pattern of a local bus service, the addresses of employees in offices, factories and shops in the town, and so on.

Use your results to plot a series of lines on a base-map, one for each catchment or circulation area identified. These lines will not of course coincide, for some shops and services have a much wider service area than others. Nor will there be a sharp break between the sphere of influence of one town and that of a neighbouring town. The best one can do is to attempt to draw an approximate 'average' line. The spheres of influence of Stanhope in Durham are shown on figure 4.10.

6 Local historical geography can be investigated by examining the changing population composition of a village over the last 100 years. You would need to obtain census data for your chosen area for 1881 and 1981, and for a few intervening periods, say every twenty years. Enumeration district boundaries must be recorded to see if any changes have taken place. Parish boundaries must also be checked when consulting the baptism, marriage and burial registers for the census years.

Your data could be summarised in map, graph and table form. This would enable you to comment on observed trends, giving reasons *why* changes have occurred.

4.3 Agriculture and land use

Data sources

1 Land use maps Two series of land use maps have been produced for England and Wales. The first was that of Dudley Stamp, mapped in the 1930s on base-maps at a scale of 6 inches to 1 mile and using fifteen categories of land use. These original maps can be read at the London School of Economics (LSE). The maps were then *published* at a scale of 1 inch to 1 mile using seven categories of land use, and can be seen at the LSE and some major libraries.

The Second Land Utilisation Survey, supervised by Alice Coleman, was completed in the early 1960s. This was recorded on 6 inches to 1 mile base-maps using 256 categories. The maps can be seen at the Land Use Research Unit, Kings College, The Strand, London WC2R 2LS. Some were published at the scale of 2½ inches to 1 mile using 70 categories of land use. These can be purchased from Kings College but cover only 15% of the area of England and Wales. From the same place can be purchased the excellent Field Mapping Manual (1981) which gives details of how to map land uses, including those of urban areas.

In 1981 three maps were published by the Land Use Research Unit, covering the whole of England and Wales. They were devised by Alice Coleman and show five landscape types. Again, they can be obtained from Kings College, London.

2 Government data sources The following statistical publications can be purchased from HMSO:

a The summary 'Agricultural Statistics', giving details of crop and livestock distributions, farm sizes and employment.

b 'Farm Classification in England and Wales' which helps summarise the characteristics of farming in a region.

c 'Farm Incomes in England and Wales', giving general details about farm incomes for different parts of the country.

d The most detailed data sources, giving annual information about crops and livestock, are the Agricultural Returns. These date back reliably to 1856, and are published in tables at the scale of individual parishes (not farms). For dates up to 1968 these tables can be read at the Public Record Office, Ruskin Avenue, Kew, Richmond, Surrey. After 1968 consult the Agricultural Census and Surveys Branch, Ministry of Agriculture, Fisheries and Food, Government Buildings, Epsom Road, Guildford, Surrey. Local offices of the same ministry may also hold data for your area. Although detailed, problems do arise when comparing the Returns for different years because of changes in parish boundaries and variations in the classification of land uses over time.

Project suggestions

Three types of agricultural land use project are particularly successful:

1 Those which examine changes in land use over a period of time. Here you should choose a relatively small area, although a single farm is usually too limited in size to be worthwhile. The most useful data source will be the Agricultural Returns. The problem will be finding out *why* land uses have changed and for this you will need a carefully worded questionnaire given to local farmers.

2 Those which examine the influence of urban settlements on agricultural land use. For example, do areas around towns have more intensive land use (possibly reflecting higher land values), are allotments a significant component of the agricultural scene (fig. 4.11), and are trespassing and vandalism a major problem for farmers?

Fig 4.11 Urban influence on agriculture: allotments

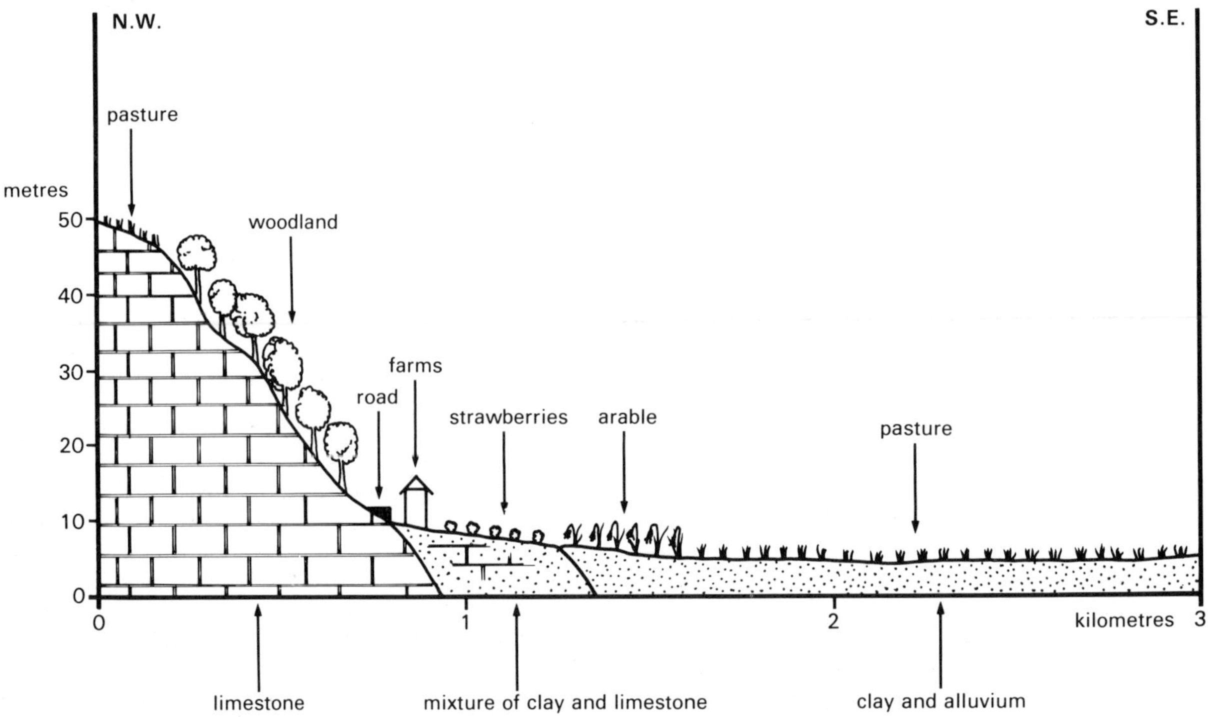

Fig 4.12 A transect down the Mendips near Shepton Mallet, Somerset, showing land uses and soil types

Fig 4.13 Land use conflict: a flooded gravel working becomes useless and dangerous land

3 Those which correlate land use with such elements as distance from markets, soils and aspect. Figure 4.12 was taken from a project which examined the impact of soil type, slope angle and distance from the farm on land use.

4 Apart from agricultural land use studies, projects examining land use conflicts such as quarrying (fig. 4.13), mining, reservoir construction and afforestation schemes are often worthwhile. For example, if you were to look at the impact of a new reservoir on an area the data sources you might use would include:
   a Questionnaires and interviews applied to local residents, planning officials and the Water Authority.
   b Published documents (including those from the local authority and Water Authority) relating to the reservoir.
   c Measurements of the local climate around the water body to see if this had changed.
   d Network analysis of the roads and paths in the area to determine whether or not local accessibility had suffered as a result of the reservoir construction (pages 109–15).

   The main aim of such a project would be to assess the relative merits and limitations of the reservoir as far as the local area and region is concerned.

4.4 Industry and commerce

Data sources

There are a large number of data sources available for studies of industry and commerce. These include the following:

1 'Kelly's Directories' and 'Kelly's Tradefinders' (fig. 4.14). Although these ceased publication in 1974, they give the addresses of firms and shops in a settlement and their function. They date back to 1845 and may be looked at in local libraries.

2 'Post Office Directories' also list information about industries and trades. They have been published for London since 1800.

3 'Yellow Pages Telephone Directories' list the telephone numbers and addresses of all businesses in a Post Office Telephone Area. They are classified by function and are thus a quick source of reference if we wish to find, for example, the addresses of greengrocers in an area. Unfortunately, not all establishments choose to enter their name in the Yellow Pages, so they are not completely reliable.

4 'Thomson Local Directories' are now available for most areas of Britain. Firms are listed by function and their addresses and telephone numbers are given. Again, not all firms choose to be entered.

TRADEFINDER. 345

CHINA & GLASSWARE SHOPS.

Head Philip (Windsor) Ltd. 17 High st. Tel 62653

PRATT & LESLIE JONES LTD,
The Token House, 26 High st & 2 & 3 Peascod st. Telephone, Windsor 63263

CHIROPODISTS.

Amdor H. LCh, HChD, SRCh, 122 St Leonard's rd. Tel 64170

Coutts Miss M. L. MChS, 11 Trinity pl. Tel 64461

Krieger H. MChS, 47a Thames av

CINEMAS.

A.B.C. Cinema, 59/60 Thames st

CINEMATOGRAPH ENGINEERS.

Specto Ltd, Vale rd. Tel 67297

CINEMATOGRAPH FILM PROJECTIONISTS & OPERATORS.

Golden Films, Stewart ho, 23 Frances rd. Tel 69566

CIVIL ENGINEERS—CONSULTING.

See Consulting Engineers.

CLEANING MATERIALS SUPPLIERS.

L.J. Supply Co (suppliers to industry, local authorities, catering trades &c), 28a Albany road, Old Windsor. Telephone No 65015 & 51688

CLINICS.

Family Planning Clinic, Kipling Memorial building, Alma rd. Tel 68111

Infant Welfare Centre, Smiths la

CLOCK REPAIRERS.

See Watch & Clock Makers & Repairers.

CLOTHING MANUFRS.

Gilbert & Davan Ltd, 8 & 10 William st. Tel 63945

CLUBS.

See Societies, Associations & Clubs.

COACH BUILDERS.

Windsor Coach Works Ltd

WINDSOR COACH WORKS LTD.
ALL TYPES OF COMMERCIAL VEHICLE BODY BUILDING AND BODY REPAIRS
Arch 35
WESTERN REGION RAILWAY GOODS YARD off GOSWELL ROAD
WINDSOR 67886

COAL & COKE MERCHANTS.

Cade Joseph & Co. Ltd. 59 Victoria st

Cadeheat, Station approach, Datchet

ETON WICK COAL CO.
43 Eton Wick rd. Eton. Telephone No. Windsor 63777

FINCH H. & SON,
6 Sherbourne drive. Tel Windsor 62526

King Bros. (Fuel Merchants) Ltd. 70 High st, Eton. Tel 62628

COFFEE MERCHANTS.

Importers, 11 Peascod st

Rombouts (G.B.) Ltd, 1a Thames av. Tel 67212

Fig 4.14 Sample page from a Kelly's Tradefinder Directory

5 Publications by individual authorities, such as the British Steel Corporation or the National Coal Board.

Project suggestions

Popular and successful types of project related to this theme are as follows:

1 A study of the distribution of industry, offices or shop types in a settlement or region. To do this you would need to mark the distribution of the phenomenon under study on a map of the town or area. This pattern could then be analysed using nearest neighbour analysis (see page 105). In the case of shop types you could supplement your study by asking a certain number of customers where they live and mapping their answers on a ray diagram (page 83). This would give you an indication of the range (the distance customers are prepared to travel to use a service) of the shop type. Similarly you could find out the places of residence of the employees of the industries or offices under study.

Figure 4.16 shows some of the results of a project which examined the distribution of bookshops and second-hand furniture shops in Oxford.

2 A study of the location and linkages between related activities in an industrial area or trading estate (fig. 4.15). This sort of project might involve the use of questionnaires designed to find out from the managers of such firms *why* locational decisions were made and how they view the location today. It is usually best to avoid studies of just *one* factory, office or supermarket unless their impact is very great.

Fig 4.15 Factory units on an industrial estate

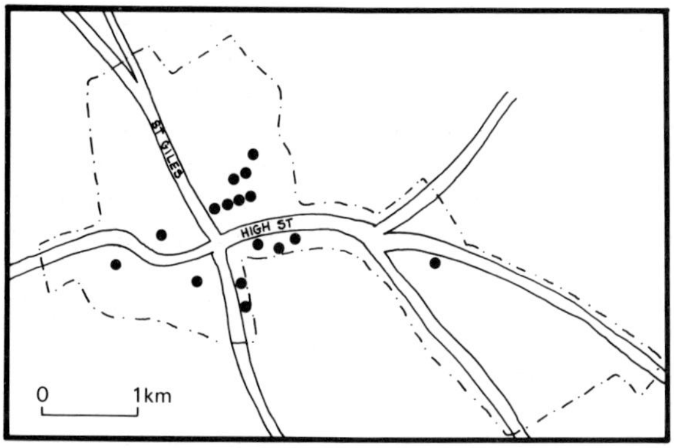

Map showing the distribution of book shops

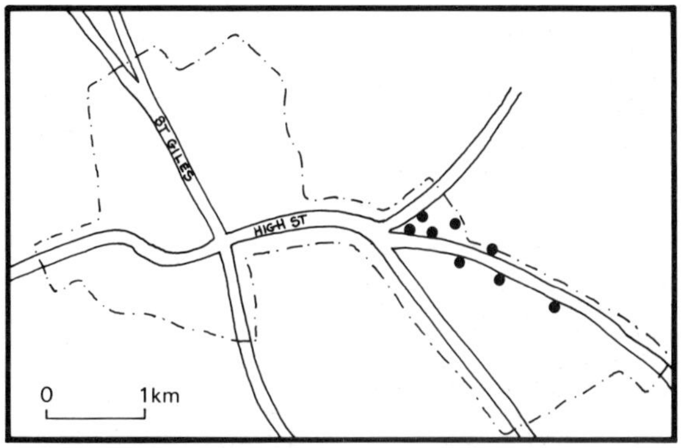

Map showing the distribution of second-hand furniture shops

Nearest Neighbour Index for book shops = **0.63**

Nearest Neighbour Index for second-hand furniture shops = **0.41**

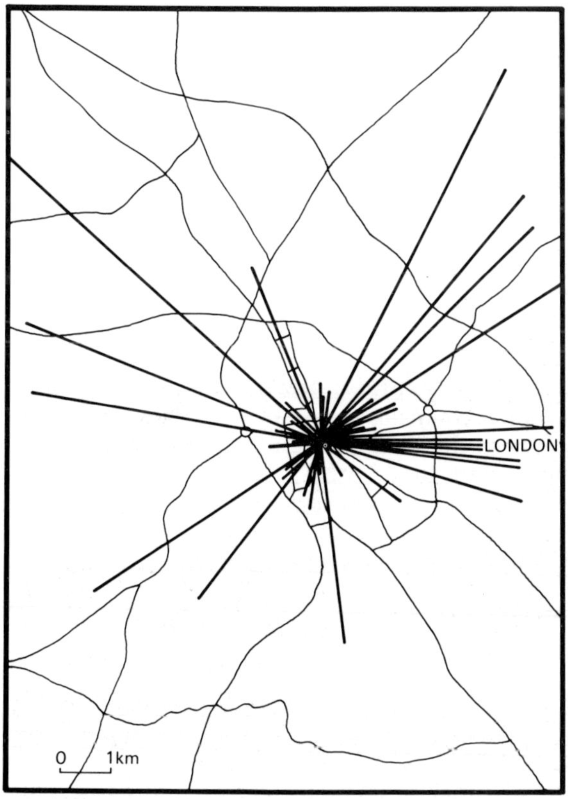

Location of residence of 40 people visiting book shops

Location of residence of 40 people visiting second-hand furniture shops

Fig 4.16 The results of a project which examined the distribution of contrasting shop types in Oxford

4.5 Recreation and tourism

Data sources

Official data sources on local recreation and tourist activity can be found, but much depends on the area concerned. Local planning authorities may be able to help you and where a specific organisation is in control of the area or facility they can often provide accurate data. Such organisations include the National Trust, the Forestry Commission, regional Water Boards, regional Sports Councils, the Countryside Commission, the Youth Hostels Association, the Nature Conservancy, the British Tourist Authority and the National Parks Commission.

Much of the data, however, will have to be collected by you. Some of the methods for doing this are described in the following project suggestions.

Project suggestions

Two types of project are particularly valuable on this theme:

1 A study of patterns of recreational activity. You could investigate the location of recreational facilities within an area and supplement this with an analysis of the people who use those facilities. You will find that questionnaires are a very useful approach in this type of project. If you word them carefully and administer them to a fair sample of people you can discover how far people come to use the facility, how long they spend there, why they choose that particular facility, how often they use it, and any other details which are important to your own particular study. You might also carry out a series of counts of the number of users of the facility at different times of the day, week or year.

Your observations can be summarised in maps, diagrams and words, enabling you to evaluate the importance of the facility to the area. Figure 4.17 shows the results of a survey of recreational activity on Hampstead Heath in London.

2 A study of the environmental impact of tourist or recreational activity and the conflicts it causes. This involves weighing up the 'pros' and 'cons' of tourist activity using, for example, questionnaires, interviews with local residents, and an analysis of hotel registers to find out the number of visitors and the seasonal distribution of their arrival. (Some hotel managers will, of course, be unwilling to let you see their register.)

4.6 Transport and trade

One of the most profound influences on our rapidly changing society has been the development of the various transport media. Faster trains, new road systems, expanding airports and larger sea-going vessels have done much to bring people closer together and expand the scope of trade. At the same time many new problems have been created: the closure of unprofitable railway lines has increased the isolation of remoter rural settlements, motorways have damaged farms and the quality of life of the people who live close to them, and so on. These are the reasons why the subject of transport and trade can be a profitable one for geographical investigation.

Data sources

Possible data sources for projects on transport and trade are as follows:

1 **HMSO publications** (mainly of use as background data only):

'National Travel Surveys', which give details about the relative popularity of different forms of transport in the UK.

The 'Digest of Port Statistics', giving details about the volume and type of trade handled by ports in Britain.

The 'Census of Distribution and Other Services', which gives general statistics about the number of shops, their turnover and staff, for urban and rural districts in Britain.

The Civil Aviation Authority's 'Monthly Statistics', giving information about the number of aircraft, passengers and volume of goods trade using British airports.

2 **Non-HMSO publications:**

The 'Annual Report of British Rail', giving information about the volume of rail traffic and its profitability.

Timetables of British Rail, local bus companies, shipping companies and airways. These can show how well different parts of a region are served by public transport.

Local authority traffic data and plans, often giving details of traffic flows for individual settlements which are not obtainable elsewhere.

Local newspapers, which may publish details of transport development schemes of interest to the geographer.

3 **Traffic counts** It is possible for you to do a count of traffic flows yourself. The most important points to decide beforehand are:

a Where are you going to do the count? Normally you will need several people helping you, stationed at key junctions.

b Between which times of day are you going to do the count and on which days of the week?

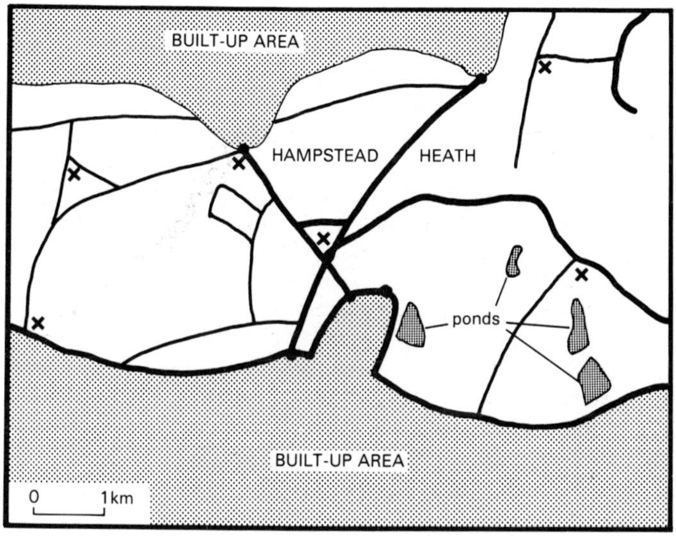

Map of Hampstead Heath. Survey sites marked with a cross

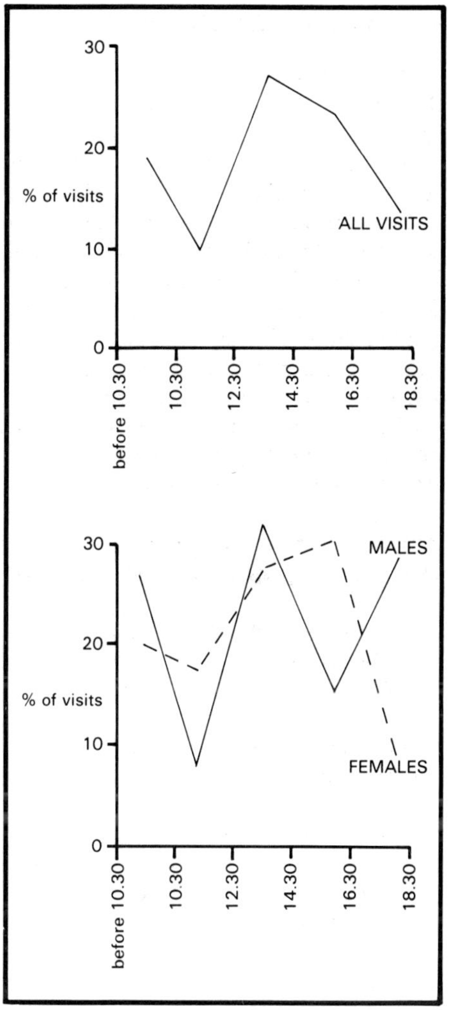

Time of arrival of visits to Hampstead Heath

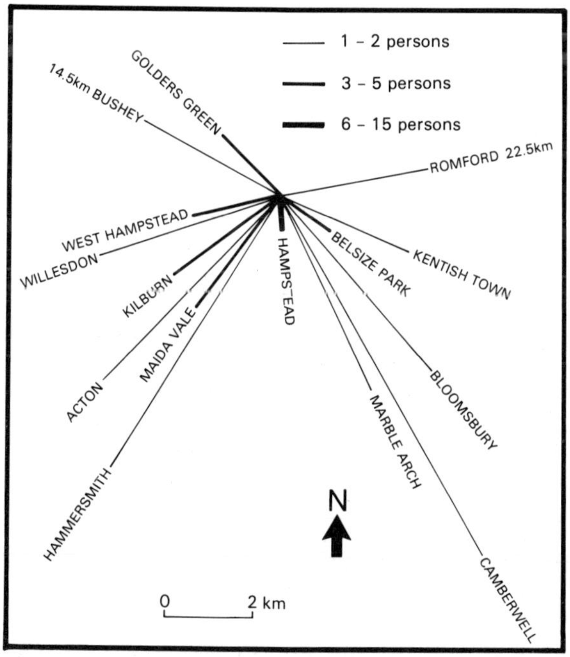

Origin of visits to Hampstead Heath

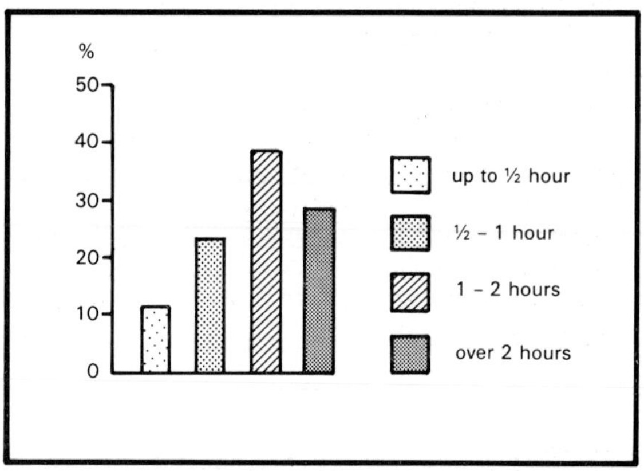

Duration of visits to Hampstead Heath

Fig 4.17 The results of a survey of recreation patterns on Hampstead Heath, London

Fig 4.18 Motorways have a great impact on the people living nearby

c Are you going to count all traffic passing a point irrespective of direction, or divide this into a two-way flow?
d Are you going to 'weight' vehicles? Some scoring system could be devised: for example, 1 point for a bicycle, 2 for a motorbike, 3 for a car, 4 for a van, 5 for a lorry or bus.

Traffic flows can be summarised in flow-line maps, which are described on page 83. Transport networks can be analysed using 'network analysis', fully described on pages 109–15.

Project suggestions

There are many possible projects related to this theme. Three successful types are as follows:

1 An analysis of changing traffic flows over time, be it a day, week or year. Primary data obtained from traffic counts will be of considerable use here. You must attempt to give *reasons* for the pattern you observe.
2 An analysis of the impact of a transport development such as a new bridge, motorway (fig. 4.18), bypass or airport. Under this heading you might investigate the impact of the development on volumes of traffic, on average journey times, and on the local environment. You will probably need to use a combination of published data sources and your own fieldwork in the form of traffic counts and questionnaires to local residents.
3 An analysis of the changing role of particular transport facilities such as an area of dockland or a canal system (fig. 4.19). Try to establish how volumes of traffic or trade have changed over time, how the nature of the individual media has changed (e.g. has the average size of the ships increased?), and how the nature of the goods carried on the system has altered. Attempt to explain why these changes have taken place, with reference to new transport technologies, competition from other systems, and the changing pattern of industrial supply and demand.

4.7 Catalogues of data sources

There are many catalogues of data sources available. From Her Majesty's Stationery Office (49 High Holborn, WC1V 6HB) can be obtained the 'Sectional Lists of Government Publications' which give titles in print divided up topically (population, transport, etc.).

Fig 4.19 The Manchester Ship Canal: transport media have to adapt to modern requirements

Part B

Processing the information

Collecting the information you require for your project will often have taken up a considerable amount of time. Once you are certain that you have all the data you need, you can begin to sift through your material to decide which of it needs processing and in what form.

The way you process your data will depend upon what you are trying to show or prove in your project. Maps, graphs, tables and statistical calculations can summarise data far more effectively than words alone, but they must be very carefully chosen. You must decide whether the material is relevant to your hypothesis or study, and if so, which method is the most appropriate one for displaying the data.

This part of the book looks at three broad strategies for processing your information: cartography, statistical methods and spatial analysis. The limitations as well as the merits of each of the techniques are discussed in order to enable you to make the most appropriate choice of method.

5 Cartography

Once data has been collected as part of a survey or project it will often be worth summarising it in some form. A good way of doing this is in a map or diagram, for relationships between large numbers of figures can be more fully appreciated if portrayed in this way. This chapter discusses some of the cartographic techniques available.

5.1 Graphs

1 Line graphs

These are used for portraying the relationship between two variables such as temperature and altitude (fig. 5.1). They are well known, but there are a few important rules which should be adhered to:

a Usually one of the two variables causes the other to change rather than vice versa. Thus if you were plotting soil depth against slope angle it is obviously the angle of the slope which causes the soil depth to change rather than the other way round. In this case slope angle is called the 'independent' variable and goes on the horizontal or 'x'-axis while soil depth is the 'dependent' variable and goes on the vertical or 'y'-axis. However, one exception to this rule is that altitude is usually plotted on the vertical axis, irrespective of whether it is the dependent or the independent variable. Thus in figure 5.1 altitude is plotted on the y-axis although it is clearly independent of temperature.

b Your axes should always start at zero.

c Always mark on the axes what the variables are.

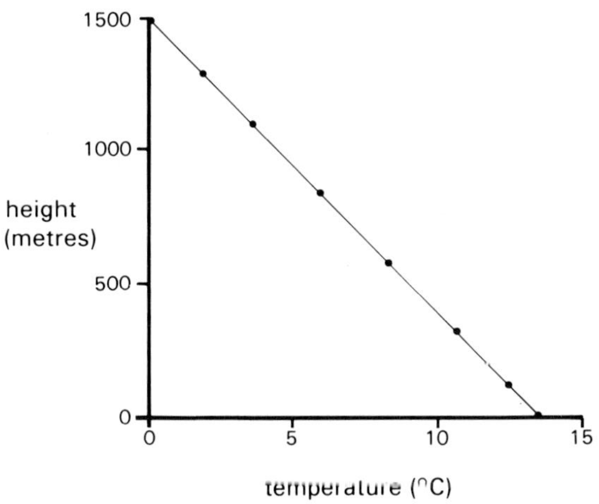

Fig 5.1 Line graph plotting temperature against altitude

d Be aware of the fact that the scales employed on the axes will very much determine the visual impression given by the graph. Figure 5.2 shows the same set of figures plotted on two graphs with axes of different scales. Note the difference in the resultant appearances of the two graphs.

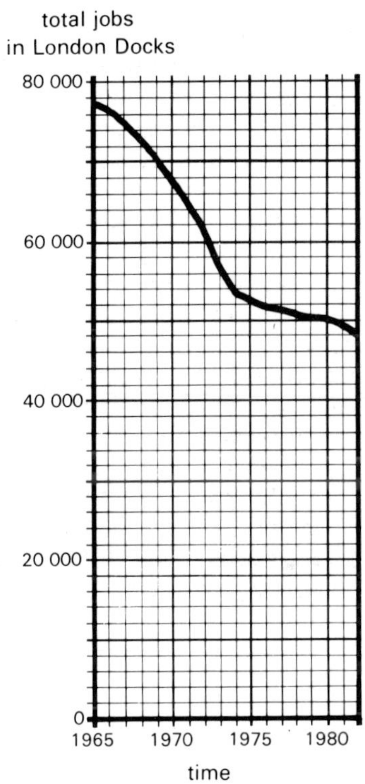

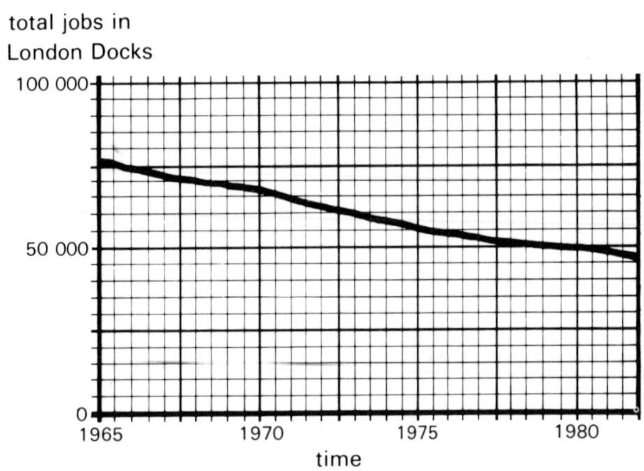

Fig 5.2 Line graphs showing the decline in the number of jobs in London Docks, 1965–82

2 Bar graphs

These are used for portraying information where one variable has a quantitative value and the other does not. Bars are drawn proportional in height to the value they are representing.

Divided bar graphs (fig. 5.3) are useful for showing two or more quantitative variables on the same graph. The one rule is that where there is a clear pattern the *largest* division of the bar should be on the bottom and the *smallest* on the top.

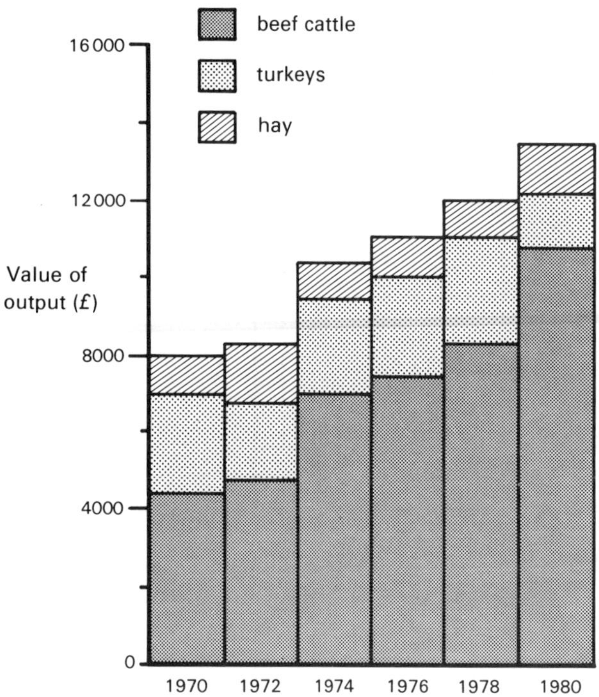

Fig 5.3 Divided bar graph showing the changing agricultural output of Forest Edge Farm, Hampshire, 1970–80

Population pyramids (fig. 5.4) are another form of bar graph. The vertical axis shows age groups in five-year intervals. The horizontal axis represents the actual number of people in each of these age groups. The whole graph is divided into two, males on the left and females on the right.

3 Circular graphs

These are used for portraying a variable which is continuous over time, such as temperature data. On a normal line graph there is a false break: it starts at one end on January 1st and ends at the other on December 31st. On a circular graph there is no such break (fig. 5.5).

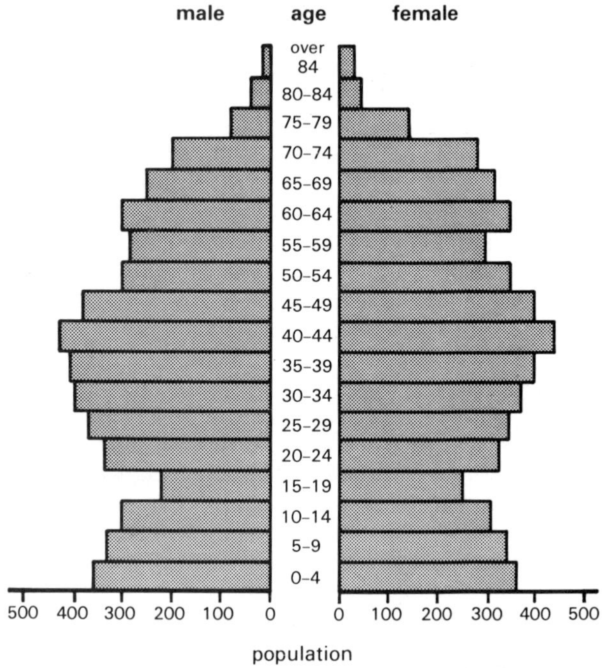

Fig 5.4 Population pyramid for the town of Ashdown, Sussex, 1982

Circular graphs are easily drawn. There are two axes – the circumference of the circle and the radius. Values (e.g. temperature) increase radially outwards. The circumference is normally time (e.g. months of the year). This is divided into 360 degrees, so a month would be 30 degrees of circumference (360 ÷ 12). It is thus a straight-line graph stretched out and bent round.

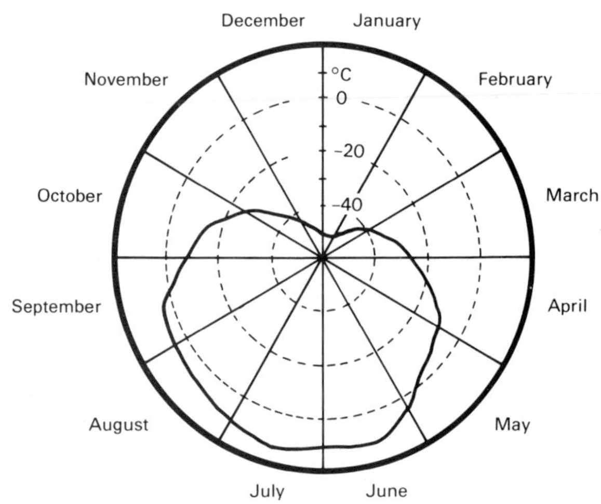

Fig 5.5 Circular graph showing temperatures at Verkhoyansk, USSR

The only major problem with circular graphs is that the easily appreciated rise and fall of a normal graph is replaced by the idea of a line moving away from or towards the centre of a circle and considerable practice is needed before its message can be appreciated in detail.

4 Logarithmic graphs

Logarithmic graphs are of two types – those where both axes are drawn logarithmically (called 'log-log graphs', figure 3.30) and those where only one axis (the vertical axis) is logarithmic ('semi-log graphs'). Figure 5.6 (b) shows a semi-log graph. So far as the horizontal axis is concerned all is quite normal. But on the vertical scale the numbers are not spaced evenly: the interval between 20 and 30 is slightly less than that between 10 and 20 and the same amount separates 200 and 300 higher up the scale. Numbers are regularly 'bunched' and each of these bunchings is called a 'cycle'. The top and bottom of each cycle must be 10 or some decimal or multiple of 10 such as 0.1, 100, 1000 or 10 000. In each successive cycle the values are 10 times greater than those of the cycle below.

Logarithmic graphs have two merits:

a It is possible to represent a very great range of data on one piece of graph paper. If you had to plot such values as 1, 3, 12 and 120 000 this would be impossible on a normal graph, possible on a logarithmic graph.

Output (tonnes per annum) from two factories 1961–1981

	1961	1971	1981
X	10	20	40
Y	50	100	200

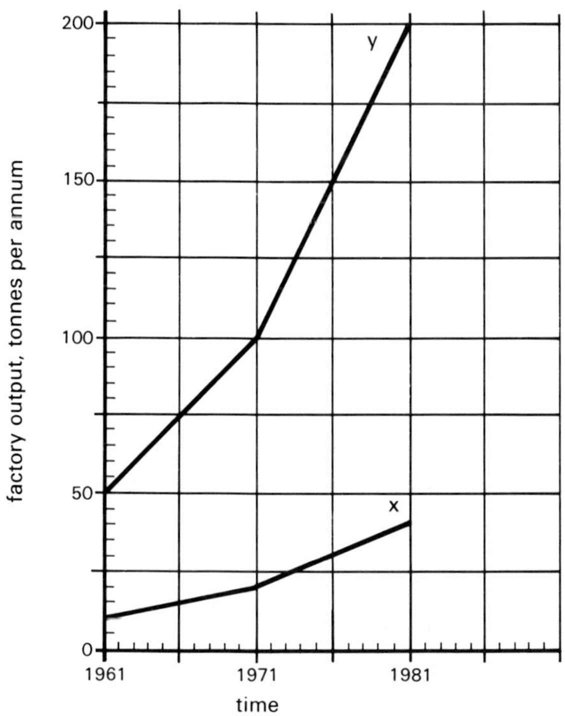

(a) Simple line graph

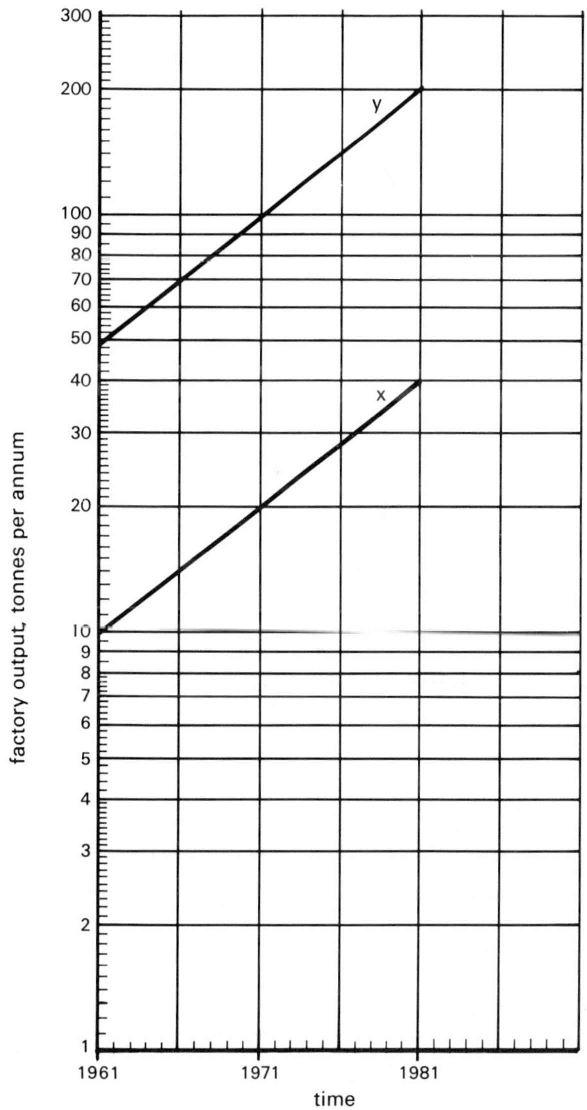

(b) Semi-log graph

Fig 5.6 Data plotted on two different types of graph

b Equal rates of change are shown by lines of equal slope. Compare figures 5.6 (a) and (b). Figure 5.6 (a) shows a normal line graph illustrating the output from two factories over 20 years. Both factories doubled their output every 10 years (equal rates of change) but the slopes of their lines are different because factory Y has higher absolute output levels. In figure 5.6 (b) both appear as a straight line. Logarithmic graphs are therefore used for plotting rates of change.

5 Scatter graphs

These are used when one wishes to investigate the relationship between *two* variables and one has data for *many* places. For example, in figure 5.7 we are investigating the relationship between population size and the number of services offered in all the settlements of a region. We could also use this type of graph to see if there was any relationship between birth rate and standard of living in 100 countries of the world, or between precipitation and discharge in 20 rivers.

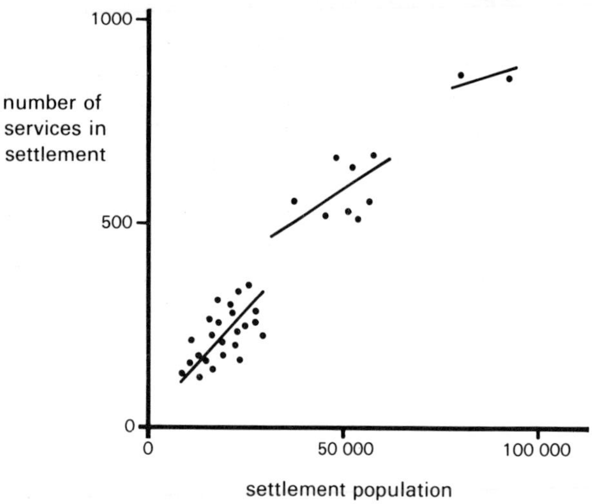

Fig 5.7 A scatter graph. All the settlements in a region were examined and the population size was plotted against the number of services in each.

The pattern of the scatter tells one about the relationship. In figure 5.7 there is a positive correlation (as one value goes up the other goes up) and there appear to be three main groups (similar places?). Lines are drawn on the graph which show the general trend of the dots. Three are drawn on figure 5.7. There should be an equal number of dots above and below the line. These are called 'best-fit' lines.

6 Divided circles (pie graphs)

These are used for portraying a quantity (such as the population of a country) which can be divided into component parts (such as different ethnic groups). A circle is drawn to represent the total quantity and it is divided into segments proportional in size to the components (fig. 5.8). It is possible to make the size of the circle itself proportional to the total quantity it represents. The method for doing this is described on page 81.

Population composition by ethnic origin San Fernando, Trinidad

1	2	3	4
Ethnic origin	Total population	%	% of 360°
Negro	18 784	47.2	169.9
East Indian	10 296	25.8	92.9
White	1 306	3.3	11.9
Mixed	8 283	20.8	74.9
Other	1 161	2.9	10.4
Total	39 830	100	360

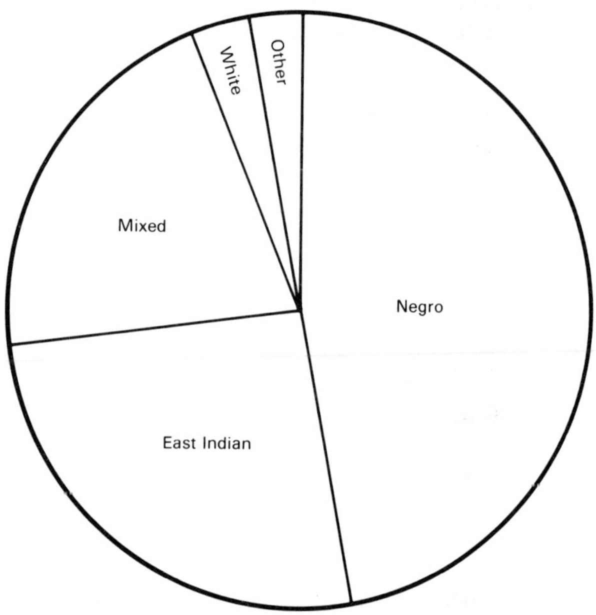

Fig 5.8 Divided circle (pie graph), with calculations, showing the ethnic composition of San Fernando, Trinidad

The method for dividing the circle is as follows:

a Draw the circle, proportional in area to the total quantity to be represented or not, as you wish.

b Tabulate the component values which will form the segments of the circle (columns 1 and 2 in the table in figure 5.8). Convert these into a percentage of the whole (column 3).

c Calculate the angle which corresponds to this percentage of 360° (column 4).

d Draw a vertical line from the centre of the circle to the top of the circumference.

e Draw in segments, measuring the angles calculated in step (c). Start from the vertical line and work in a clockwise direction. Draw the segments in order of size (largest first, etc.).

f Different segments may be shaded differently and numbers or words written in.

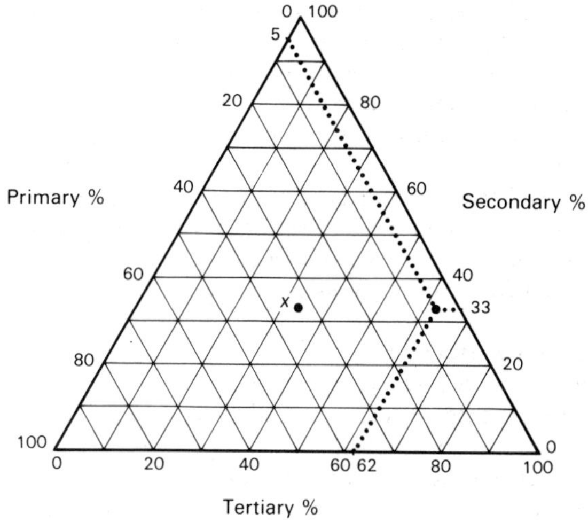

Fig 5.9 Triangular graph showing the employment structure of a town

7 Triangular graphs

Triangular graphs, sometimes known as 'ternary diagrams', are graphs with three axes instead of two, taking the form of an equilateral triangle (fig. 5.9). The important features are:

a Each axis is divided into 100, representing percentages.

b From each axis lines are drawn at an angle of 60° to carry the values across the graph (fig. 5.10).

c The data used must be in the form of three components, each component representing a percentage value and the three percentage values adding up to 100%. One could plot, for example, the particle constituents of a soil sample:

Particle	**per cent**
Silt	10
Clay	25
Sand	65
	100

. . . or the employment structure of a town:

Employment	**per cent**
Primary	5
Secondary	33
Tertiary	62
	100

These latter figures have been plotted on figure 5.9. Dotted lines show the way that the values are carried across the graph until they meet at one point. The position of this point indicates the relative dominance of each of the three components. If exactly one-third of workers were engaged in each of the three employment types the point would be in the middle of the triangle, marked *x* on figure 5.9. Care must be taken when plotting and interpreting such a graph, for it can be confusing at first acquaintance.

Triangular graphs can be used for plotting any data which can be conveniently divided into three portions, for example agricultural land use (arable, pastoral, other) or age structure (0–25, 26–50, over 50). Their main value arises when data for several places is plotted on one graph, the relative position of the points giving a quick visual impression of the relative dominance of one component or another.

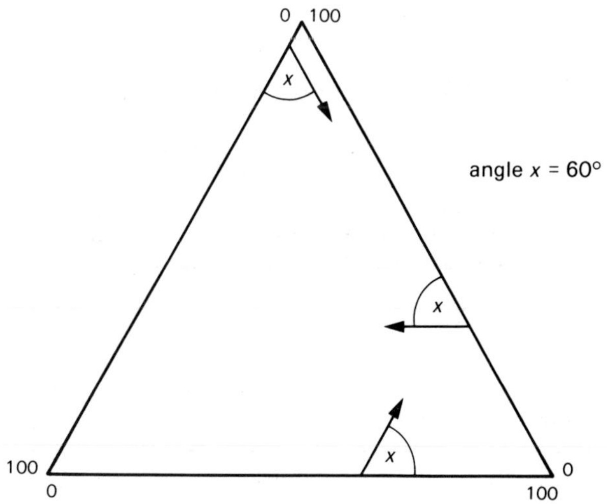

Fig 5.10 Diagram of a triangular graph showing the way in which the values are carried across the graph

5.2 Isoline maps

Isolines are lines on a map which join points of equal value. We are all familiar with isolines – contours on a relief map, isotherms of temperature and isobars of pressure, for example.

Isolines can only be used when the variable to be plotted changes in a fairly gradual way across space and where plenty of data is available. If the spatial distribution is disjointed and data is not detailed enough too much guesswork is involved in the drawing of the isolines.

The method is simple:

1 Plot the data on a map as a series of points with accompanying values.
2 Decide on the interval you want between your isolines. If this is too small there will be many isolines and the map will look cluttered; too great and the map becomes too generalised to be useful.
3 Draw in the isolines. You must stick to your chosen interval: all isolines should have the same interval between them. There is a good deal of personal judgement involved here. Knowledge of things like 'isobars tend to be circular around low pressure systems but straighter between the fronts' is useful.
4 The space between different value isolines can be shaded or coloured. The higher the value of the isolines, the darker the shading. This is the system used for portraying relief in atlases, except that conventionally greens are used for low land, browns for high land, purple and white for very high mountains. If you do shade or colour, include a key. If you do not shade between isolines mark on their numerical value. Figure 5.11 illustrates this.

MERITS: Isolines are ideal for showing gradual changes over space and avoid the 'unreal' effect which boundary lines produce on choropleth maps (section 5.5).

LIMITATIONS: They are unsuitable for 'patchy' distributions, a large amount of data is needed for an accurate isoline map, and a good deal of personal judgement is always involved.

5.3 Dot maps

In dot mapping dots of a fixed size are given a value representing a variable such as crop yield or numbers of people, cattle, shops, etc. They are located on a base-map approximately in the known position of that phenomenon. Figure 5.12 shows the distribution of tractors in eastern Brazil in 1950 and 1960. Two scales of dot have been used, a small dot representing 100 tractors and a larger dot representing 1000 tractors.

The main steps involved in the drawing of a dot map are as follows:

1 Prepare your data. If the data you are using was gathered within enumeration districts obtain or draw a map showing the enumeration district boundaries. Remember that the larger the areal units used the less informative the map will be.
2 Find the total number of items to be shown on the map – the number of people, cattle, etc.
3 Decide on the dot value. This should be high enough to avoid excessive overcrowding of dots in areas with a high concentration of the phenomenon being mapped, and low enough to prevent areas with low concentrations of the phenomenon having

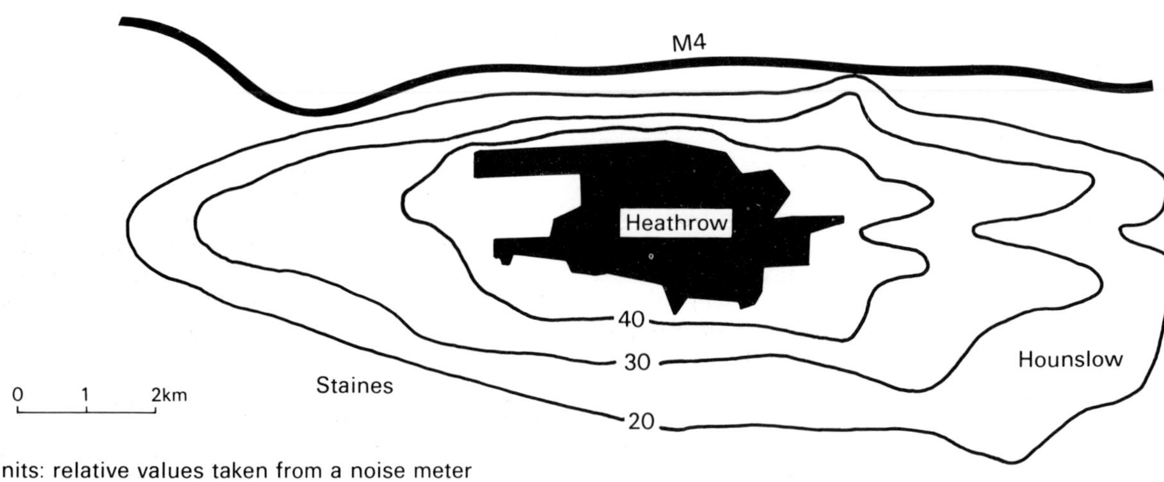

Fig 5.11 Isoline map of aircraft noise contours around Heathrow Airport, 1982

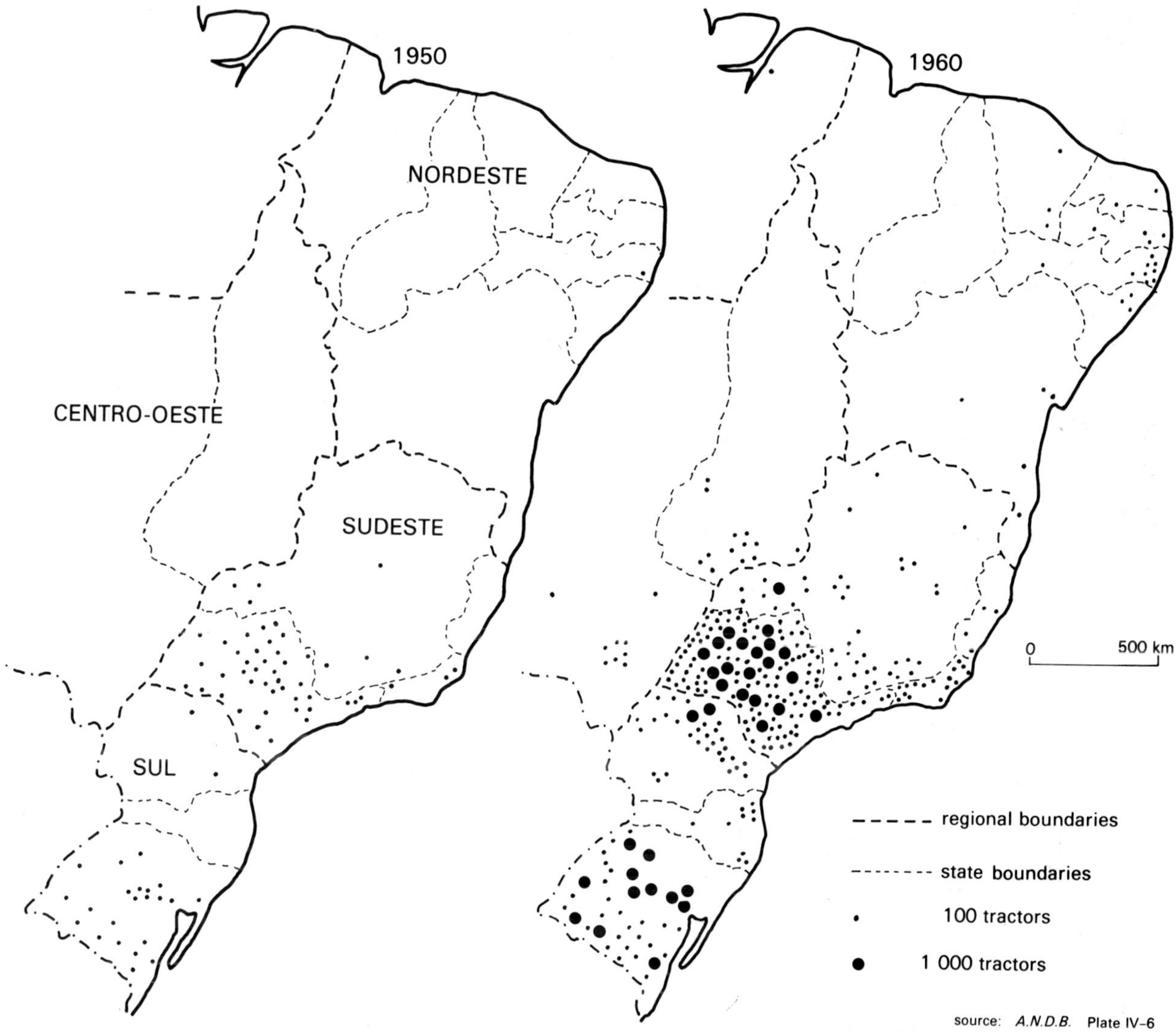

Fig 5.12 Dot maps showing the spread of tractors in eastern Brazil between 1950 and 1960

no dots at all, so giving a false impression of emptiness. It may not be possible at any one time to fulfil both these criteria and some compromise value will have to be decided on. Alternatively two, or maybe three, different value dots may be used.

4 Decide on dot size. Dots of the same value should be drawn of uniform size, this size depending on the density of dots in the area of the map with the highest density. Ideally, dots should not be too large and should not merge.

5 Dot location: dots are located on the map so as to reflect as closely as possible the distribution of the phenomenon being mapped. If the values are given for enumeration districts some dots should be placed close to the district boundaries to avoid leaving areas blank. Within the districts dots may be located subjectively over the known locations of, say, villages or farms.

6 Dots of different colours can be used. For example, if one were mapping the distribution of members of five different ethnic groups in an American city, five different dot colours could be used. The same dot value should be used for each.

MERITS: Dot maps are useful for showing the distribution of phenomena where values are known and a fairly accurate indication of their location is given. It is the only technique which gives this accurate indication of distribution.

LIMITATIONS:

a Dot maps suffer from the problem that large numbers of dots are hard to count, so that while they are very good at giving an *impression* of distribution they are less valuable if a precise estimate of the values they represent is required.

b If the dots are to be plotted in enumeration areas and nothing is known about the distribution of the phenomena *within* these areas, other methods (such as choropleth mapping) are better. In other words, one must have a large amount of initial information before one begins to draw a dot map.

5.4 Proportional symbols

Proportional symbols are simply symbols drawn on maps proportional in size to the size of the variable being represented. The symbol used can theoretically be almost anything: proportional 'men' to show military strength, proportional 'trains' to show the number of trains in an area, proportional 'factories' to represent industrial output. More common are proportional spheres and cubes, drawn three-dimensionally. But the most usual and straightforward are proportional bars and circles.

Proportional bars

The method for drawing proportional bars (fig. 5.13) is as follows:

1 Examine the data and decide on your *scale*. The length of the bar will be proportional to the value it portrays. If they are too long the bars get in each other's way, but if they are too short the difference between one bar and another is hard to see.

2 Draw your bars on a base-map, one end of the bar located next to the place to which it refers. Bars should be of uniform width – solid-looking but not

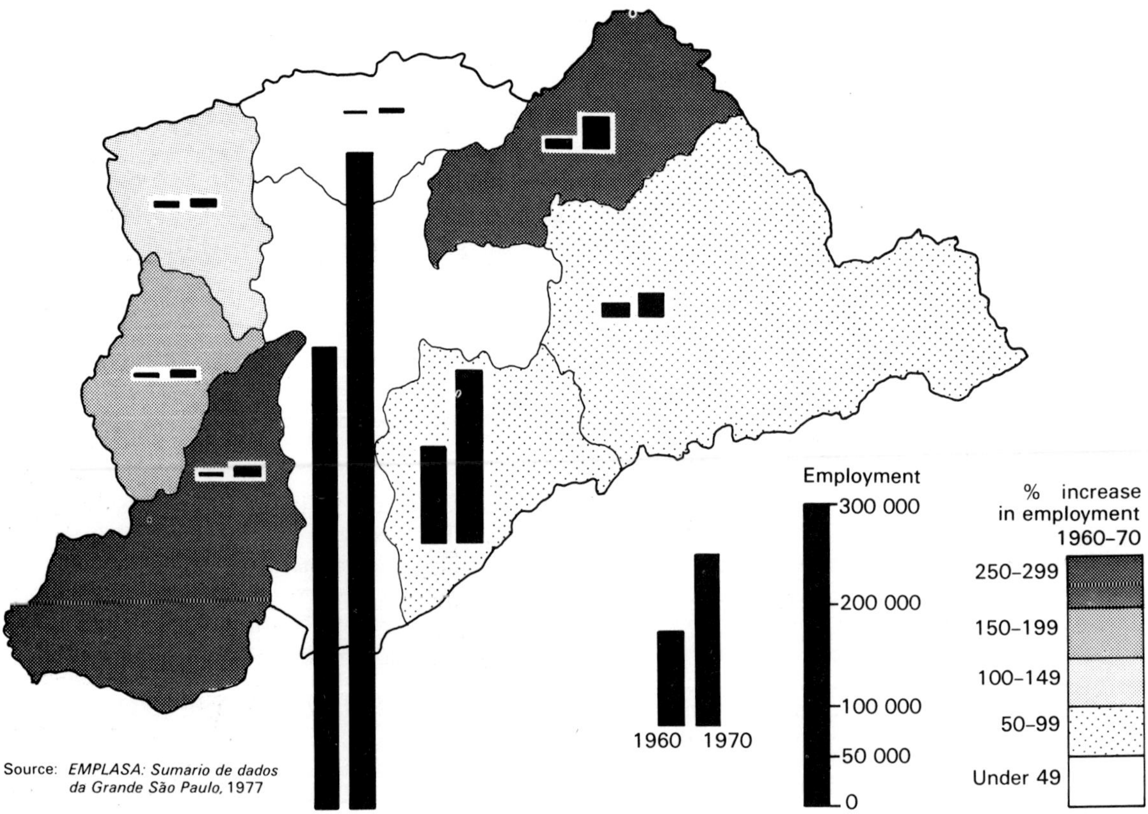

Fig 5.13 Proportional bars showing numbers employed in manufacturing in 1960 and 1970 in areas of Sao Paulo, Brazil, superimposed on a choropleth map showing the percentage increase in manufacturing employment over the period

so broad that they overlap. They may be placed vertically or horizontally.
3 Mark the scale on the map.
4 Bars may be divided (page 74).

MERITS: Bars are easy to draw and simple to estimate visually.

LIMITATIONS:
a Because bars are linear it is hard to show data which has a great range: some bars would either be tiny or huge.
b Furthermore large bars are not very attached visually to the locality they are supposed to symbolise. It is hard to get ideas of distributions from a bar map.

Proportional circles

Circles drawn proportional to the size of some given variable (fig. 5.14) are a common cartographic method. The procedure for drawing these is as follows:
1 Examine the data to be mapped and decide on a scale. If the range of values is great, convert them into their square roots. This reduces the range of the data so that, if circles are drawn proportional to these values, we avoid both excessively large and very small circles. The data is then converted into millimetres and drawn as the *radius* of the circle. If the circles are too big they will either overlap the boundaries of the area they represent or each other. If they are too small differences between them will be hard to see.
2 Draw the circles and mark the scale on the map.
3 The circles can be divided (page 76).

MERITS:
a If square roots are used a broad range of data can be shown.
b The circle is a compact shape.
c The circles can be divided.

LIMITATION: They are quite time consuming to draw.

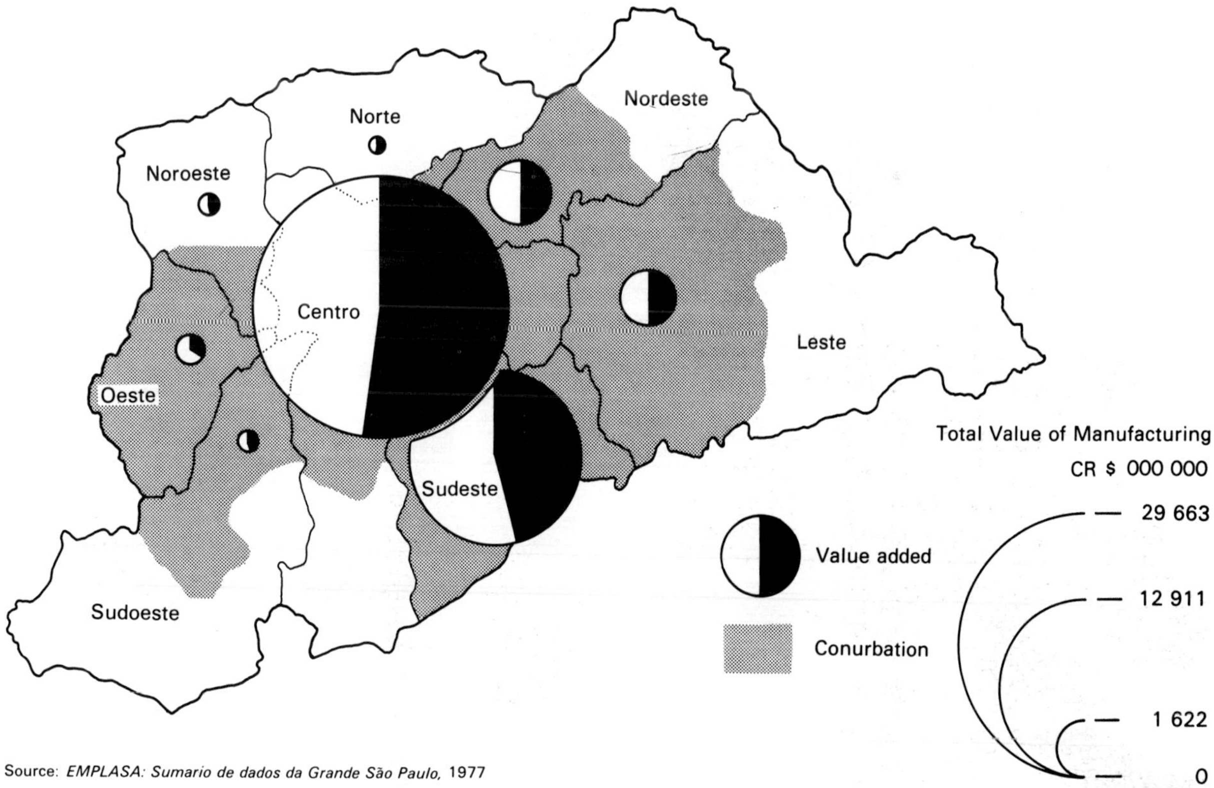

Fig 5.14 Divided proportional circles showing the total value of manufacturing and value added by the manufacturing process in different parts of Sao Paulo, Brazil

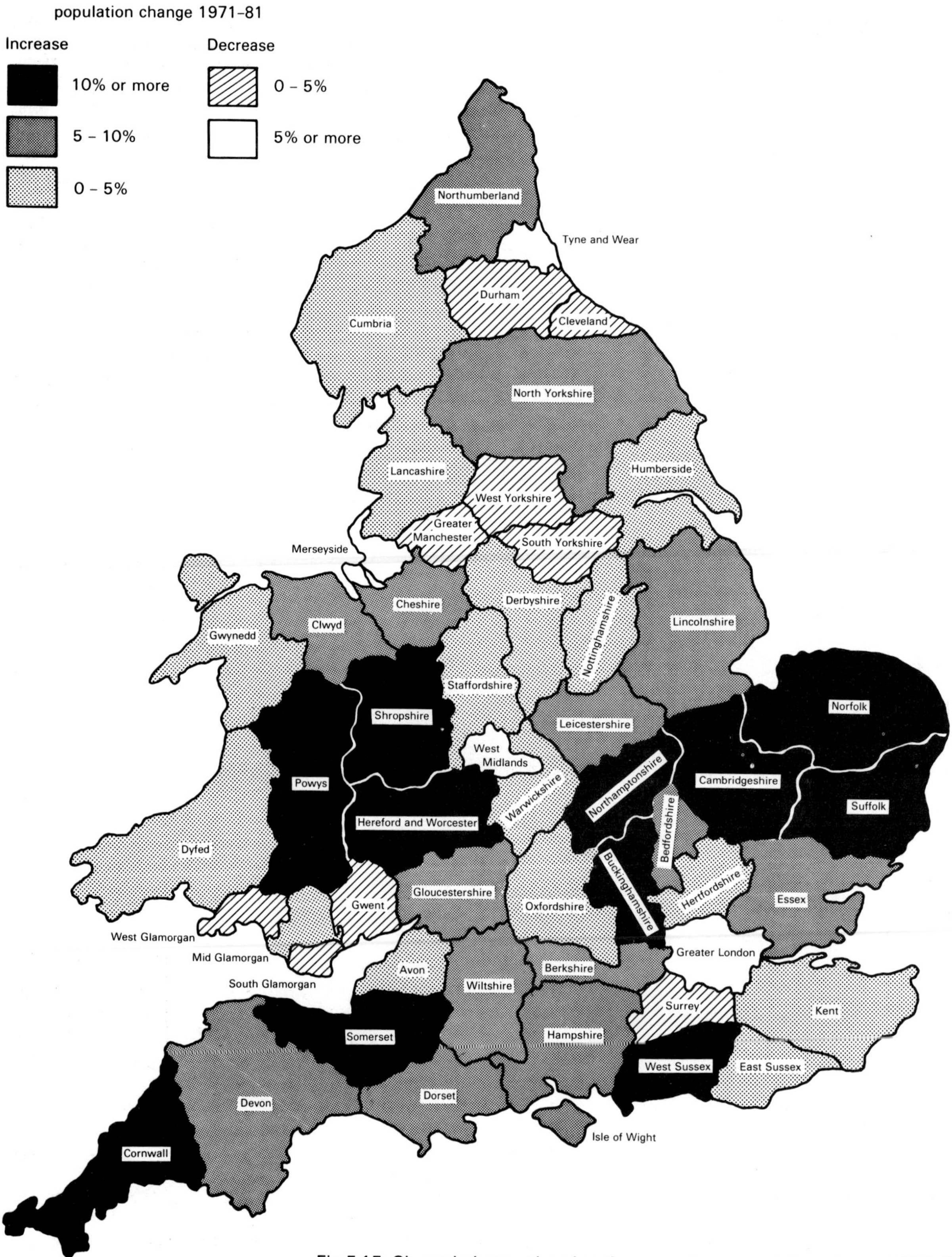

Fig 5.15 Choropleth map showing the percentage population change 1971–81 in counties of England and Wales

5.5 Choropleth maps

In choropleth or shading maps (fig. 5.15) areas are shaded according to a pre-arranged key, each shading or colour type representing a range of values. The main steps involved in the drawing of choropleth maps are as follows:

1 Obtain a base-map, with boundaries of the areal units for which you have data marked on. The smaller the areal units the more accurate the map will be.
2 Find the range of your data and devise a shading scale accordingly. For best visual results you should have no fewer than four shading types and no more than ten, depending of course on the level of detail required. Your shading should get *darker* the higher the value. This applies whether using black and white or coloured shading. If colours are used it is best to use groups of colours from the spectrum, such as:

red	high value	violet
orange	↓	blue
yellow		green
white	low value	yellow

3 Shade in the areal units and draw a key on the map.

MERITS: The choropleth shading method is easy to do and gives a good visual impression of change over space as long as a suitable shading system is used.

LIMITATIONS:
a The main limitation of the choropleth map is that it gives a false impression of abrupt change at the boundaries of areal units. This is an unavoidable problem when a technique of this type is used.
b Variations *within* areal units are concealed and for this reason smaller units are better than large.

5.6 Flow-line maps

Flow-line maps (fig. 5.16) are used for portraying movements or flows, such as traffic flows along roads or flows of migrants between countries. A line is drawn along the road, or from the country of origin to country of destination, proportional in width to the volume of the flow.

The method for constructing a flow-line map is as follows:

1 Draw a base-map, marking relevant details such as areal units or the course of a road. If the map is to represent flows along a road network mark on, in pencil, the points at which vehicles were counted and the quantity of traffic counted at those points.
2 Examine the *range* of your data and decide upon a scale. If the data range is not too great and the route density not too high the scale can be a directly proportional one, for example:

   1 mm thickness : 100 cars per hour
   2 mm thickness : 200 cars per hour, etc.

   If the range of data is large or the route density is high you will need to convert the data into a compressed form, for instance into square roots or logarithms of the quantity involved, and draw a line proportional to this. The important thing to remember is that if the flow lines are too wide they will tend to block-up and obscure the map.
3 Draw the flow lines. These may go along the actual course of the phenomenon being mapped, *or* direct from origin to destination, *or* by some other more convenient route as long as the 'tail' of the flow line begins at the flow origin and the 'nose' of the line points towards the destination.

   If the flow is a two-way movement this can be shown by dividing a flow line and shading it, one shading type representing flow in one direction, the other shading type representing the reverse flow (fig. 5.17).

5.7 Ray diagrams

Ray diagrams consist of straight lines (or 'rays') which show a movement or connection between two places. On any one base-map different coloured rays can be drawn to show different kinds of movement or types of connection between places. There are several types of ray diagram:

Desire lines

A desire line diagram shows the movement of phenomena from one place to another. Each line joins the places of origin and destination of a particular movement. Figure 5.18 shows desire lines running from the homes of people to those settlements where they go shopping for specified goods.

Wind roses

A wind rose has rays focusing on the point from which wind direction observations have been made (fig. 5.19). Each ray is proportional in length to the number of days per year that wind from the direction in which that ray is pointing has blown.

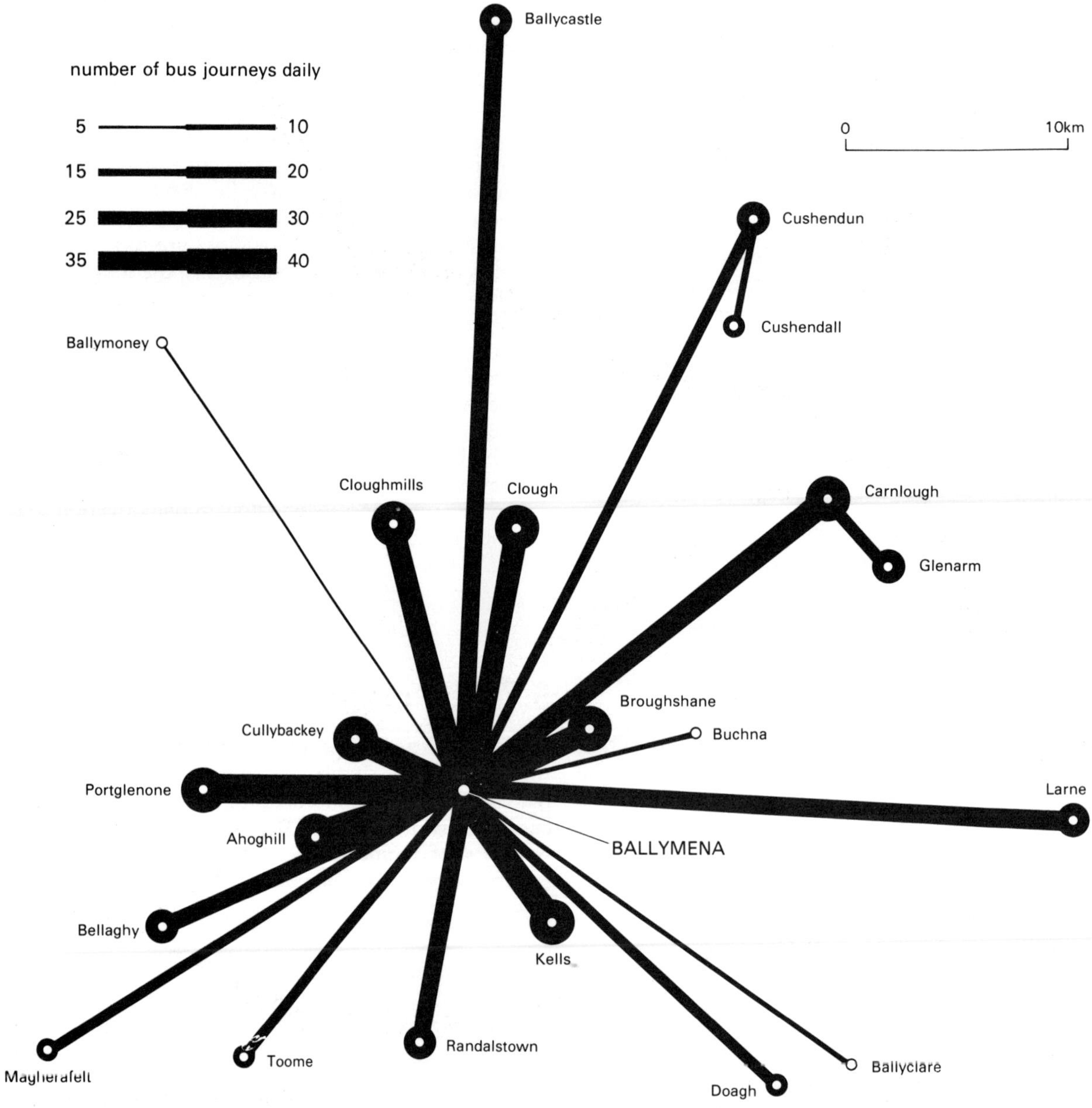

Fig 5.16 Flow-line map showing the number of daily bus journeys from Ballymena, N. Ireland, to surrounding settlements

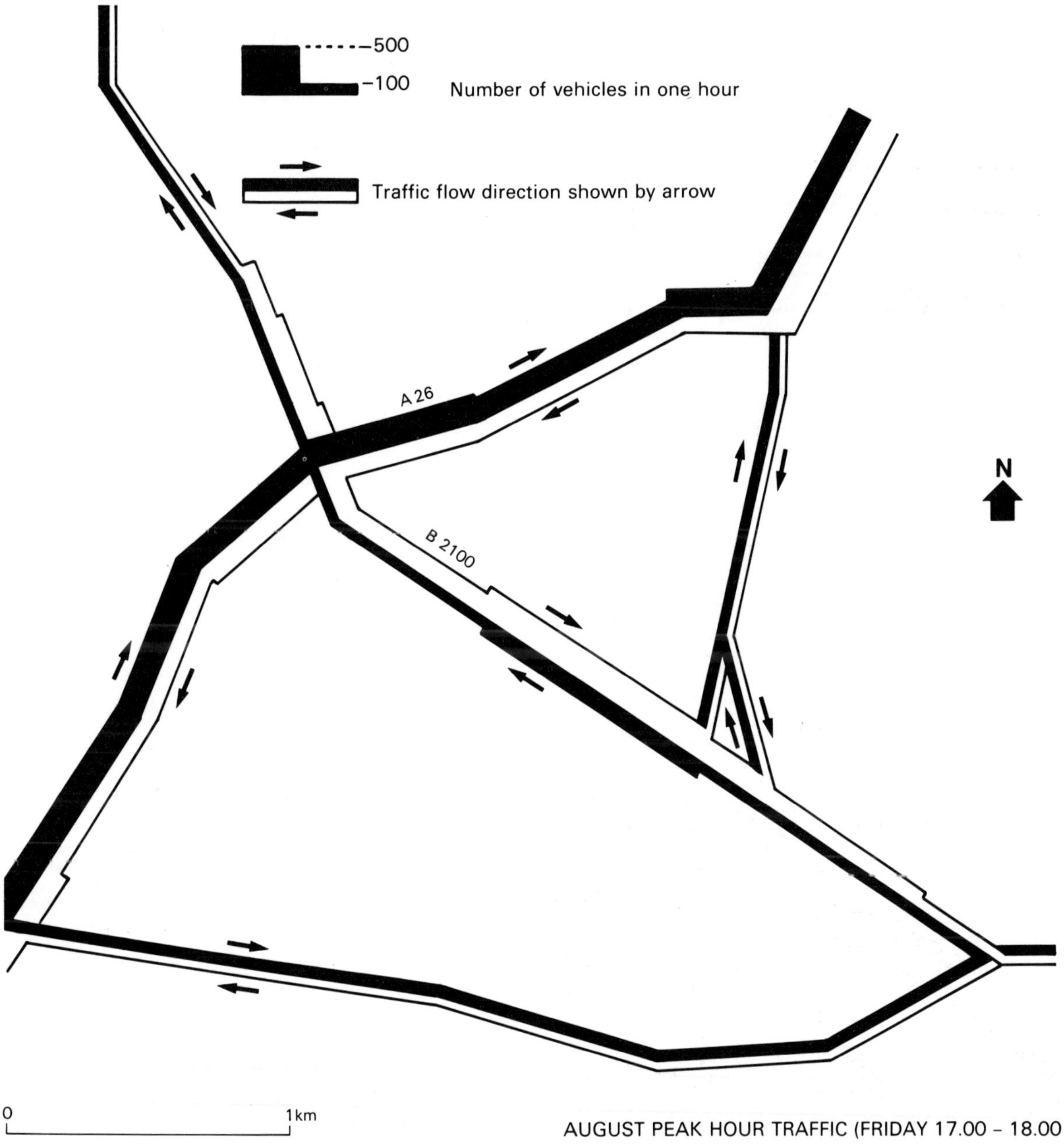

AUGUST PEAK HOUR TRAFFIC (FRIDAY 17.00 – 18.00 hrs)

Fig 5.17 Flow-line map of traffic flows in Crowborough, Sussex

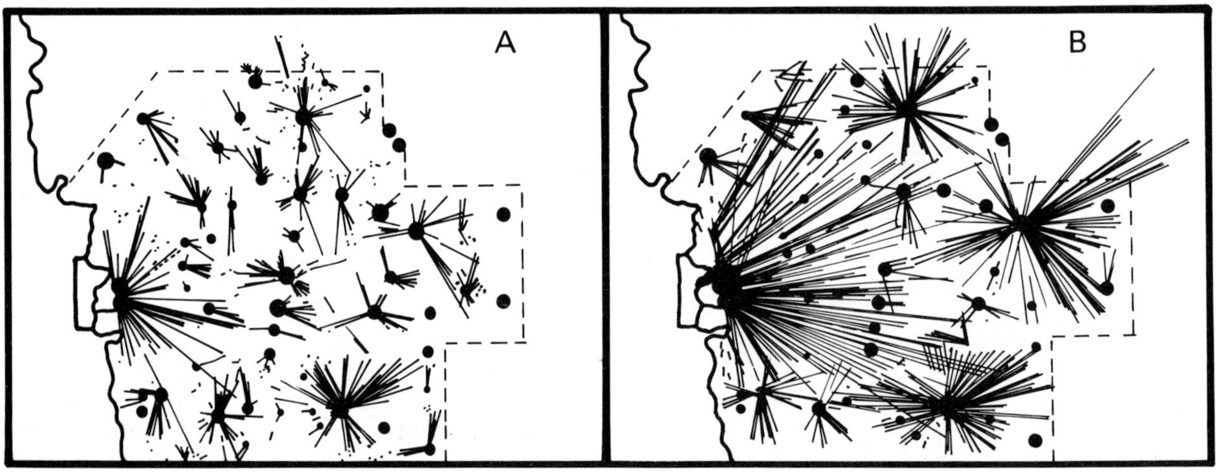

Fig 5.18 Ray diagram (desire lines) linking the homes of people in Iowa, USA, to those settlements where they purchase A) groceries and B) women's clothes

Direction of wind	N	NE	E	SE	S	SW	W	NW	Calm
Number of days per year	26	37	39	32	30	57	60	53	31

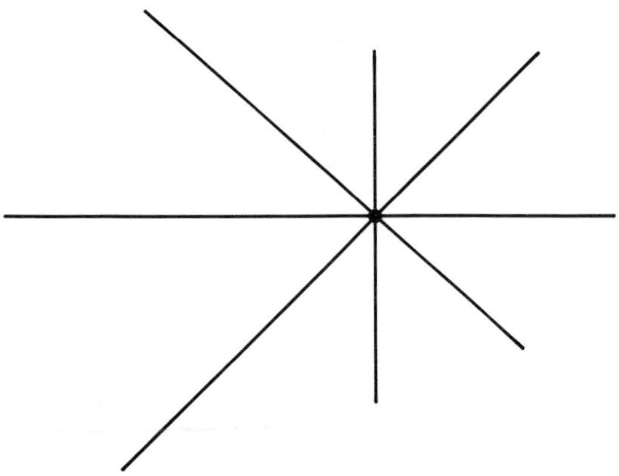

Fig 5.19 Wind rose showing wind directions recorded for one year at a school weather station in NW England. Each line is drawn proportional in length to the number of days on which wind from that direction was blowing.

Kinship ties

Figure 5.20 shows rays running from the centre of a council estate in the town of Sidcup in Kent to the homes of local relatives of fifteen people interviewed on that estate. It shows two things: the *number* of relatives those fifteen people have living in the town and the *location* of them.

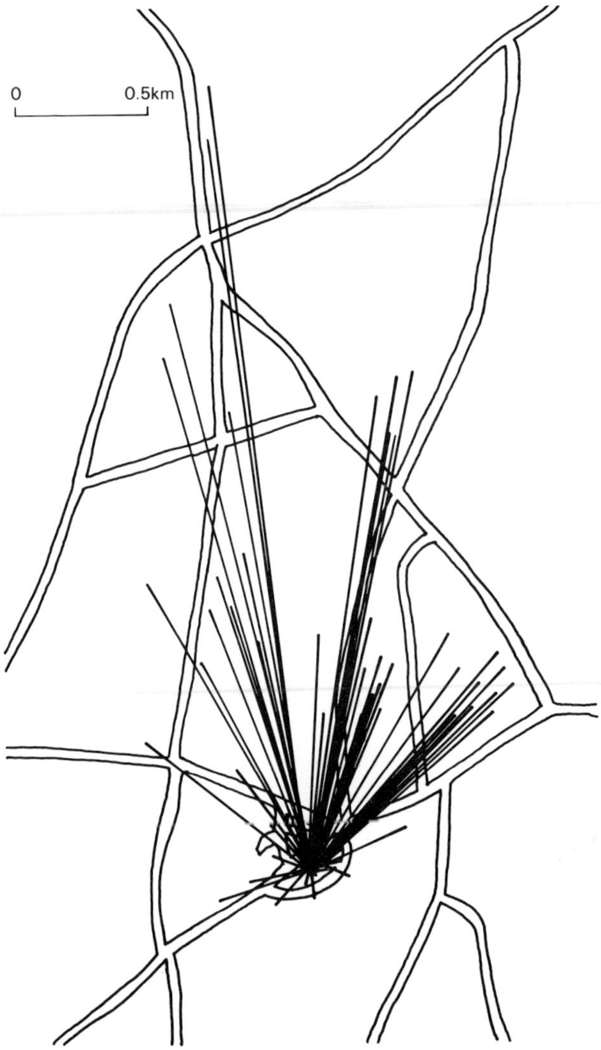

Fig 5.20 Ray diagram (kinship ties) showing the location of relatives of 15 residents of a council estate in Sidcup, Kent

5.8 The Lorenz curve and Gini coefficient

The Lorenz curve

This is a graph illustrating the concentration or diversity of activities within an area: for example, the degree of industrial specialisation or diversification of a town, region, or country, or the concentration or dispersal of immigrants within the same.

The method for drawing a Lorenz curve is as follows:

1 Obtain your data. The data given in figure 5.21 shows the number of people employed in each of fifteen categories of jobs in the Spitalfields area of East London, a region which specialises in textiles.
2 Rank your values and place them in rank order (column B).
3 Convert each value into a percentage of the total (column C).
4 Add these percentages cumulatively (column D).
5 Plot columns B and D against each other (fig. 5.22).

If the same number of people were employed in each category the plotted figures would give a straight diagonal line. The greater the deviation of the curve from the straight diagonal, the greater the degree of concentration of the place on one or a few occupations.

The Lorenz curve is valuable because it gives a visual impression of concentration. It is most useful for comparing several places whose respective Lorenz curves are plotted on one graph, or one place at several periods to show changes over time, for example, in the degree of occupational specialisation. It should be remembered that the value of the Lorenz curve depends on the quality of one's initial data. If, for example, the data used to plot figure 5.22 included some very broad occupational categories along with some rather narrow ones the Lorenz curve would be less meaningful. It must also be noted that the result is partly affected by the areal units for which the data is given. The larger the number of areal subdivisions used, the greater will be the apparent measure of concentration. For this reason one can only compare Lorenz curves for different places if the areal units within which the data was collected were the same for each place.

	A	B	C	D
Occupation	Number employed	Rank	Percentage of total	Cumulative percentage
Textiles	4000	1	68.5	68.5
Printing, paper, publishing	380	2	6.5	75.0
Distributive industries	320	3	5.5	80.5
Miscellaneous services	300	4	5.2	85.7
Public administration	150	5	2.5	88.2
Transport and communications	130	6	2.2	90.4
Construction	120	7	2.1	92.5
Insurance, banking, finance and business services	115	8	2.0	94.5
Food, drink and tobacco	110	9	1.9	96.4
Professional and scientific services	100	10	1.7	98.1
Instrument and electrical engineering	50	11	0.9	99.0
Gas, electricity and water	30	12	0.5	99.5
Other manufacturing	25	13	0.4	99.9
Coal and petroleum products, chemicals and allied products	5	14	0.05	99.95
Primary industries	5	15	0.05	100.0
	5840		100.0	

Fig 5.21 The number of people employed in each of 15 categories of jobs in the Spitalfields area of East London (column A). Columns B–D show the calculations for drawing a Lorenz curve to illustrate these figures (see figure 5.22).

The Gini coefficient

This is a means of expressing the Lorenz curve in numerical terms. It is found quite simply from the formula:

Gini coefficient =

$$\frac{\text{Area of the graph above the diagonal representing even distribution}}{\text{Area of the graph between the diagonal and the curve}}$$

The area can be easily calculated by counting boxes and portions of bisected boxes on the graph paper. The Gini coefficient varies between one (the whole distribution concentrated in only one category) and infinity (totally even distribution). The same proviso as was made for Lorenz curves about the size of areal units in which data was collected and the comparability of two places applies to the Gini coefficient.

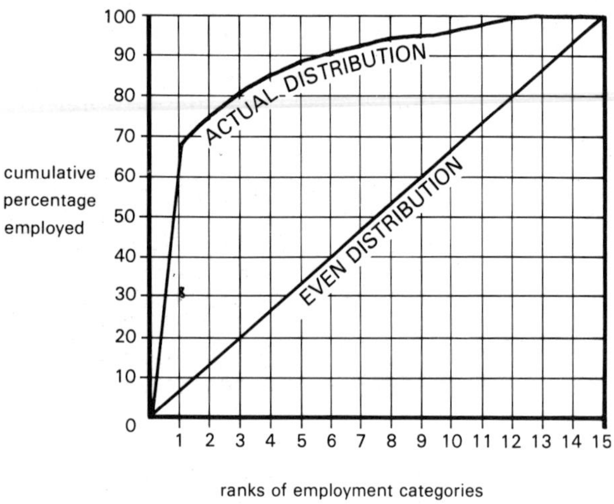

Fig 5.22 Lorenz curve showing the degree of occupational specialisation of Spitalfields, East London

The Gini coefficient for figure 5.22 is therefore as follows:

$$\text{Gini coefficient} = \frac{\text{Area above diagonal}}{\text{Area between diagonal and curve}} = \frac{75 \text{ boxes}}{57 \text{ boxes}} = 1.3$$

The Gini coefficient is useful because it is a more accurate indicator of concentration or diversity than the Lorenz curve. It is hard to visually compare two Lorenz curves which are only slightly different, but this difference will be brought out by comparing their respective Gini coefficients.

5.9 Exercises

1 With reference to figure 5.24 (barley production in the Chilterns in 1951):
  a Describe fully the cartographic method used for each of the maps.
  b Assess the relative suitability of each of the techniques used for providing information about the crop.

2 On a copy of a map of Australia, with state boundaries marked on, represent the following data using appropriate cartographic techniques:

State	Population of capital city	Population outside capital city	Total
New South Wales	3 000 000	1 900 000	4 900 000
Victoria	2 600 000	1 000 000	3 600 000
Queensland	1 000 000	900 000	1 900 000
S. Australia	900 000	400 000	1 300 000
W. Australia	900 000	300 000	1 200 000
Tasmania	200 000	300 000	500 000
Capital Territory	150 000	50 000	200 000
N. Territory	40 000	60 000	100 000

3 Figure 5.23 (a triangular graph) shows the changing population structure of a village in central Wales. Describe the changing age composition of this settlement.

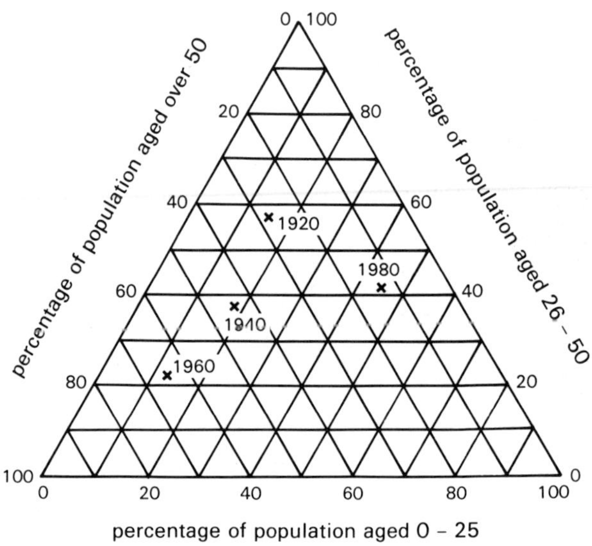

Fig 5.23 Triangular graph showing the changing population structure of a village in central Wales over time

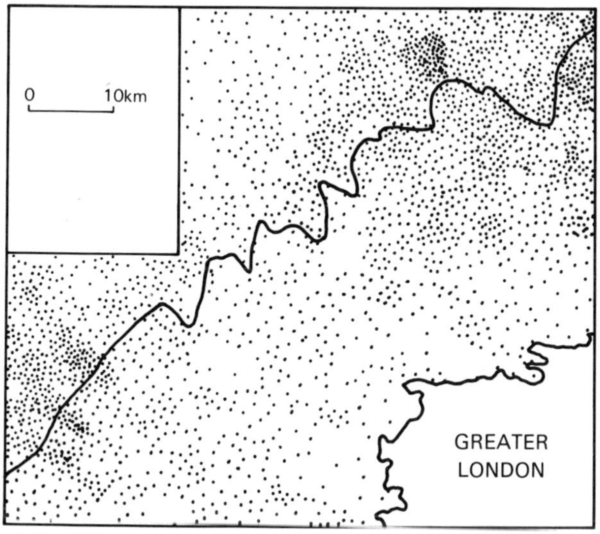

(a) each dot represents 20 acres of barley

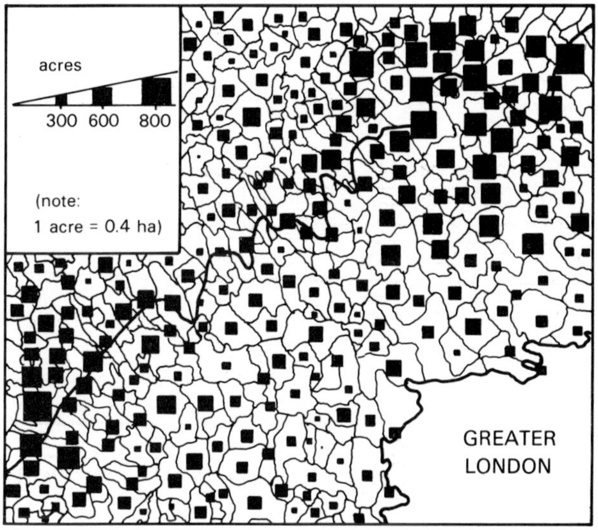

(b) the area of each square is proportional to the acreage of barley

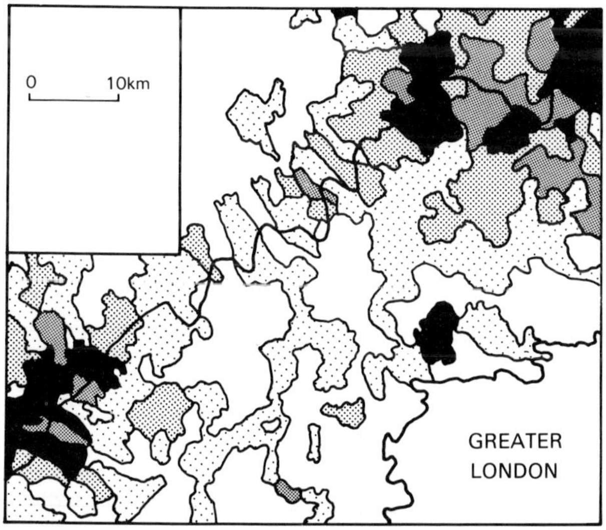

(c) percentage of agricultural land under barley

under 6 | 6–10 | 11–15 | 16–20 | over 20

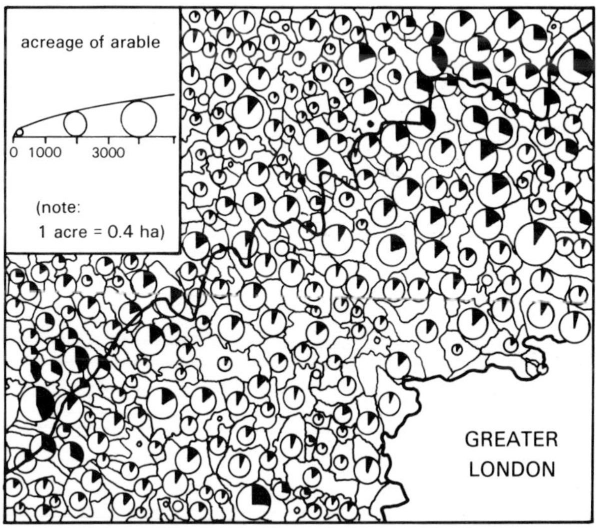

(d) the area of each sector is proportional to the acreage of barley

Fig 5.24 Four maps showing the distribution of barley production in the Chilterns in 1951

4 a Draw Lorenz curves on graph paper to represent the following data for two towns in the Midlands:

Town A	Social class	Rank	Population
	III	1	20 000
	IV	2	15 000
	II	3	12 000
	I	4	8 000
	V	5	7 000
			62 000

Town B	Social class	Rank	Population
	IV	1	9 000
	V	2	4 000
	III	3	2 000
	II	4	600
	I	5	400
			16 000

b Calculate the Gini coefficients for the two towns.
c What do your results show?

5 Represent the following data in graphical form:

Population data for a town in West Africa, 1980

Age	Male	Female
0–4	740	745
5–9	650	650
10–14	620	620
15–19	420	430
20–24	300	350
25–29	290	320
30–34	250	270
35–39	240	250
40–44	220	220
45–49	150	150
50–54	100	100
55–59	100	100
60–64	70	70
65–69	50	60
70–74	30	40
over 74	25	35

6 Statistical methods

Statistical methods are used to take the analysis of data one stage beyond that which can ever be achieved with maps and diagrams. A visual comparison of data is often a vital first step but mathematical manipulation usually gives greater precision and allows us to discover things which might otherwise go unnoticed.

Having said this, however, it is most important to be aware that statistics are merely an *aid* to analysis and no more. Too often statistical calculations are made in geographical projects without adequate justification. Before you use a statistical method it is essential to ask yourself two questions:

1 Why am I using this technique? In other words, be absolutely clear what it is you hope to prove and how this statistical method can help you do it.
2 Is my data appropriate to this particular technique? As explained below, each technique requires data to be arranged in a particular form. If it is not, the technique cannot be used. Above all, if your data is not good in the first place the use of a complex statistical technique will not help it. As the adage says: 'Rubbish in – rubbish out'.

In this section each technique is described step by step and worked examples corresponding to these steps are given.

6.1 Data summaries

If faced with a large amount of data, such as the average temperature of a place every day for two years, you will probably wish to get it into a more manageable form by summarising it. This is relatively easy to do and there are three commonly used methods:

1 The mean

The mean is what you probably know as the average and is found by adding together all the values under consideration and dividing the total by the number of values. The mean is shown by the symbol $\bar{x}$.

The mean is distorted if you have just one extreme value, which can be a problem. But it is the most widely used summary because it can be used for further mathematical processing.

EXAMPLE
Data: 3, 4, 4, 4, 6, 6, 9

Working: $\frac{3 + 4 + 4 + 4 + 6 + 6 + 9}{7} = \frac{36}{7} = 5.1$

$\bar{x} = 5.1$

2 The mode

The mode is simply the most frequently occurring event. If we are using simple numbers the mode is the most frequently occurring number. If we are looking at data on the nominal scale (grouped into categories) the mode is the most common category.

The mode is very quick to calculate but it cannot be used for further mathematical processing. It is not affected by extreme values.

EXAMPLES

Data: 3, 4, 4, 4, 6, 6, 9

mode: most frequently occurring number = 4

Data:

Land use	Hectares
Clover	10
Rye	12
Vegetables	15
Fruit	3
Wheat	29
Barley	18
Pasture	17

mode: most frequently occurring category = wheat

3 The median

The median is the central value in a series of ranked values. If there is an even number of values the median is the mid-point between the two centrally placed values.

The median is not affected by extreme values but it cannot be used for further mathematical processing.

EXAMPLES

Data: 3, 4, 4, 4, 6, 6, 9

median: central value = 4

Data: 3, 4, 4, 6, 6, 9

median = 5

6.2 Spread around the median and mean

The median, mean and mode all give us a summary value for a set of data. On their own however they give us no idea of the spread of data around the summary value; this can be misleading. For example, if we were looking at rainfall figures for a semi-arid area the data might read as follows:

Year	Rainfall (mm)
1976	0
1977	0
1978	3
1979	0
1980	97

The mean for this data (20 mm) gives an untrue picture of what really happened. There is a very great 'deviation about the mean'. The median (0 mm) is similarly misleading – the deviation is even greater.

Deviation can be measured statistically in a number of ways:

1 Spread around the median: the inter-quartile range

The inter-quartile range is a measure of the spread of values around their median. The greater the spread the higher the inter-quartile range.

METHOD

Stage 1 Place the variables in rank order, smallest first, largest last.

Stage 2 Find the 'upper quartile'. This is found by taking the 25% highest values and finding the mid-point between the lowest of these and the next lowest number.

Stage 3 Find the 'lower quartile'. This is obtained by taking the 25% lowest values and finding the mid-point between the highest of these and the next highest value.

Stage 4 Find the difference between the upper and lower quartile. This is the 'inter-quartile range', a crude index of the spread of values around the median. The higher the inter-quartile range, the greater the spread.

EXAMPLE

Monthly average temperatures (°C), Middleton, Norfolk

January	4	July	17
February	5	August	17
March	7	September	15
April	9	October	11
May	12	November	7
June	15	December	5

Ranked: 4 5 5 7 7 9 11 12 15 15 17 17

lower quartile 6 (↓ between 5 and 7); median 10 (↓ between 9 and 11); upper quartile 15 (↓ between 15 and 15)

Inter-quartile range: (15 − 6) = 9

2 Spread about the mean: the standard deviation (σ)

If we want to obtain some measure of the spread of our data around its mean we can calculate its standard deviation. It is useful to know this because obviously the greater the range or spread of the data, the less useful is the mean as a summary of it.

METHOD

Stage 1 Tabulate the values (x) and their squares (x^2). Add these values (Σx and Σx^2).

Stage 2 Find the mean of all the values of x ($\bar{x}$) and square it ($\bar{x}^2$).

Stage 3 Calculate the formula:

$$\sigma = \sqrt{\left(\frac{\Sigma x^2}{n} - \bar{x}^2\right)}$$

where σ = standard deviation
$\sqrt{\ }$ = square root of
Σ = the sum of
n = the number of values
$\bar{x}$ = the mean of the values

The higher the standard deviation, the greater the spread of data around the mean. The standard deviation is the best of the measures of this spread because it takes into account *all* the values under consideration.

EXAMPLE

Data: No. of vehicles passing a traffic count point on 10 days between 9.00 and 10.00 a.m.

Day 1	50	Day 6	70
Day 2	75	Day 7	63
Day 3	80	Day 8	42
Day 4	92	Day 9	75
Day 5	60	Day 10	82

Working:

x	x^2
50	2 500
75	5 625
80	6 400
92	8 464
60	3 600
70	4 900
63	3 969
42	1 764
75	5 625
82	6 724
$\Sigma x = 689$	$\Sigma x^2 = 49\,571$

$$\bar{x} = \frac{689}{10} = 68.9$$

$$\bar{x}^2 = (68.9)^2 = 4747.2$$

$$n = 10$$

$$\sigma = \sqrt{\left(\frac{\Sigma x^2}{n} - \bar{x}^2\right)}$$

$$= \sqrt{\left(\frac{49\,571}{10} - 4747.2\right)}$$

$$= 14.5$$

This figure, 14.5, is the standard deviation or mean deviation of the data from the mean. It is a useful figure to have because it tells us much more about the data than the mean alone can do. In our example, if we were planning a traffic management scheme to cope with the traffic on this road we would not only be interested in the average (or mean) amount of traffic which passes along that road but also in the amount of traffic on particularly busy and idle days of the week. The greater the standard deviation, the greater the difference in volume between these high and low periods. If the standard deviation was very high we might wish to install traffic lights which only operated at the busiest times of the day or week.

3 Spread about the mean: the coefficient of variation (V)

The coefficient of variation is simply the standard deviation expressed as a percentage of the mean. The higher the coefficient, the greater the spread of data.

The coefficient of variation can be calculated from the formula:

$$V = \frac{\sigma}{\bar{x}} \times 100$$

where V = coefficient of variation
σ = standard deviation
$\bar{x}$ = the mean of the values

Being in percentage form, coefficients of variation from different sets of data are more readily comparable. For this reason the coefficient of variation is often used to plot variability on a map (fig. 6.1).

EXAMPLE

Traffic count data (from previous example):

$\bar{x} = 68.9$

$\sigma = 14.5$

$$V = \frac{\sigma}{\bar{x}} \times 100$$

$$= \frac{14.5}{68.9} \times 100$$

$$= 21\%$$

6.3 The normal distribution curve

The 'frequency distribution' of data is the form taken by data when plotted on a graph, with data groupings on one axis and the frequency of occurrence of the data within each of these groupings on the other.

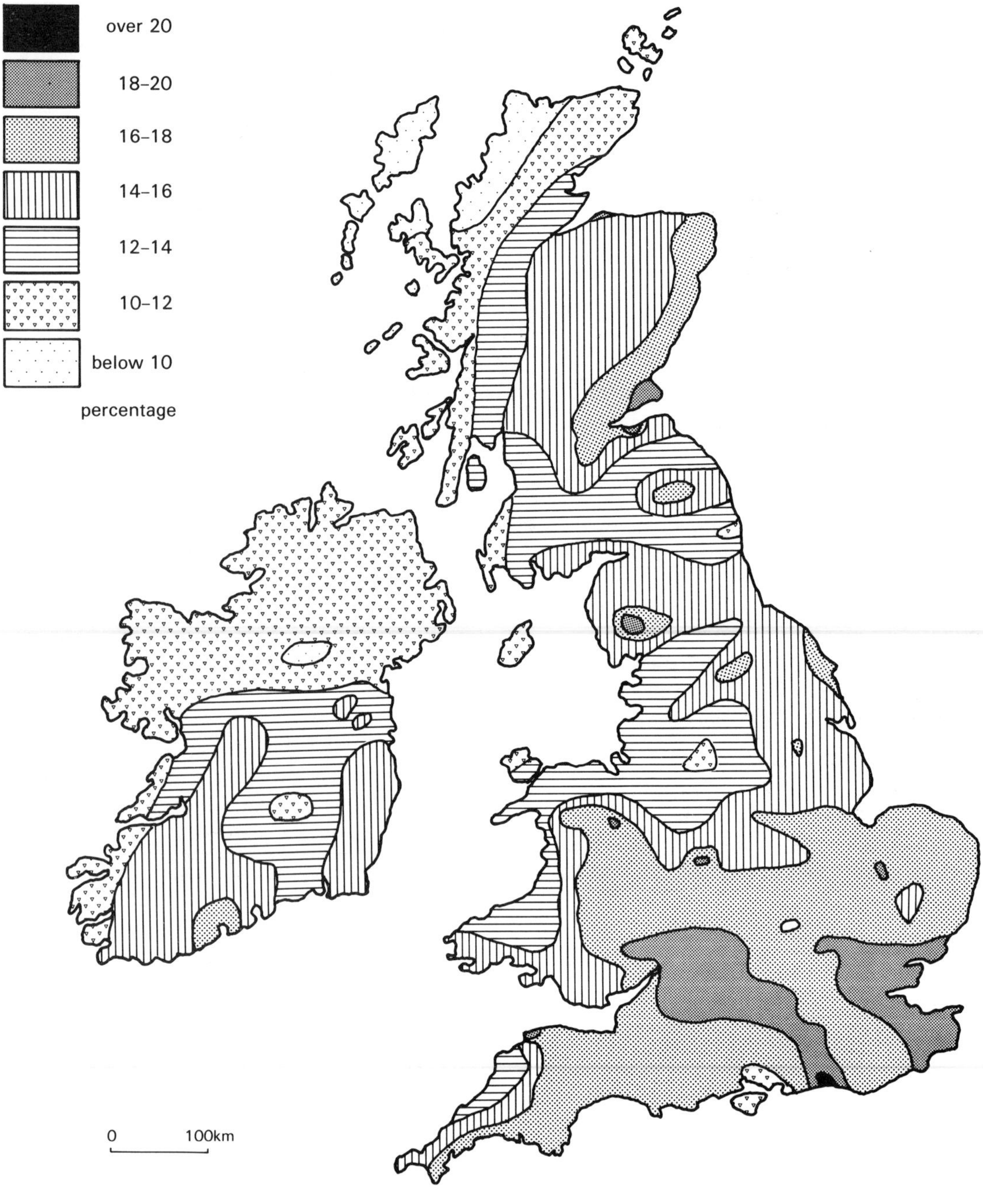

Fig 6.1 The coefficient of variation (V) of annual rainfall over the British Isles, 1901–30. The higher the coefficient, the more variable the rainfall from year to year.

When plotted in grouped form the result is a series of bars known as a 'histogram' (fig. 6.2). The mid-points of the tops of the bars can be joined to give a curve, a summary of the frequency distribution of the data. It is found that for many types of data this frequency distribution curve takes on a regular bell-shape (fig. 6.3). This is known as a 'normal distribution curve' and the frequency distribution of the data is regular about its mean.

A normal distribution curve arises where the factor measured (for example, rainfall) is influenced by two sets of forces:

1 *Dominant* and *constant* forces which produce the tendency to cluster around the mean. In the case

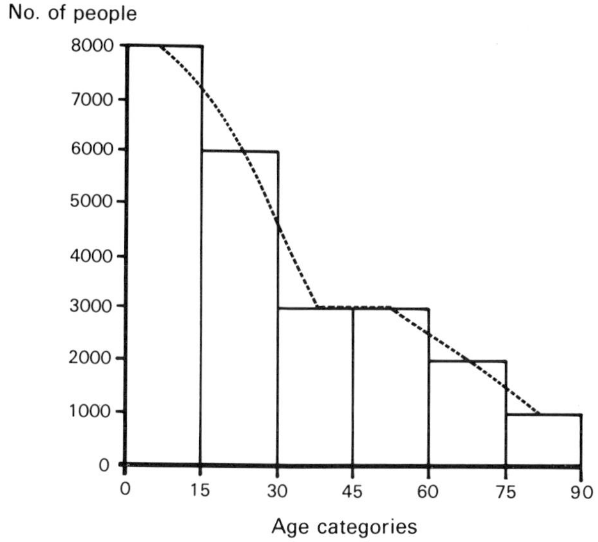

Fig 6.2 Histogram of population age categories for Sherborne, Dorset

of rainfall these would include such things as latitude, longitude, altitude and proximity to the sea.
2 *Secondary* and more *random* forces producing dispersion around the mean, such as the irregular occurrence of depressions, anticyclones and different air masses.

Normal distribution curves are often found, owing to the common existence of these two sets of forces. Plotting only a small sample of the data (for example, only 30 years of annual rainfall) may not produce a very regular normal distribution. But the normal curve may emerge if the data is plotted for a much larger sample.

The normal distribution curve is symmetrical about the mean. Apart from this stipulation, other components of its shape do not matter; all the curves in figure 6.3 are normal curves.

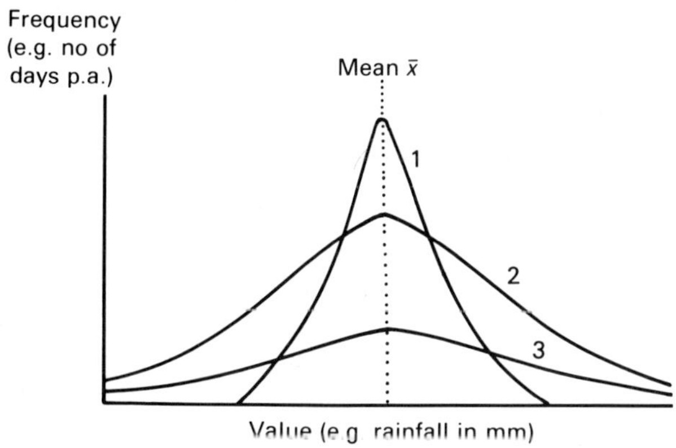

Fig 6.3 Normal distribution curves

6.4 Probability

Everyone knows that if a coin is flipped a large number of times the 'mathematical probability' is that it will land as 'heads' 50% of the time and 'tails' 50% of the time. It may not *actually* do this but will almost certainly approximate to this pattern. Thus mathematical probability tells us what will happen in theoretical rather than real terms.

Probability can be expressed in two ways:

1 As a percentage: for example – 'there is a 50% probability of this happening'.
2 As a fraction of one: the percentage probability divided by 100. For example – 'there is a 0.5 probability of this happening'.

Probability and the normal curve

Certain useful probability statements can be made about data which, when plotted, forms a normal distribution curve. These statements all use the standard deviation of the data which is a measure of the dispersion of the values about the mean (page 92).

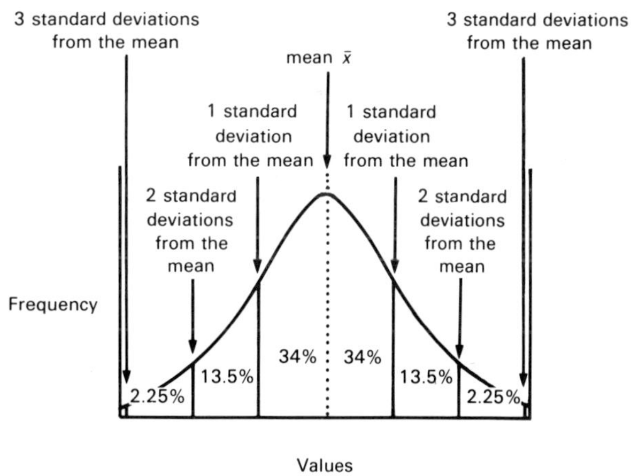

Fig 6.4 A normal distribution curve showing the proportion of values (in percentages) which fall within a given distance (measured in standard deviations) from the mean ($\bar{x}$)

The simplest rule is that if a set of data has a near-normal distribution, then just over 68% of the values will fall within one standard deviation (plus and minus) of the mean, just over 95% of the values will fall within two standard deviations of the mean and

99.5% will fall within three standard deviations of the mean. These statements are summarised in figure 6.4.

The statements can also be expressed in probability terms:
There is a 0.68 or 68% probability that a value will be within one standard deviation of the mean.
There is a 0.95 or 95% probability that a value will be within two standard deviations of the mean.
There is a 0.995 or 99.5% probability that a value will be within three standard deviations of the mean.

Or expressed another way:
There is a 0.32 or 32% probability that a value will be more than one standard deviation from the mean.
There is a 0.05 or 5% probability that a value will be more than two standard deviations from the mean.
There is a 0.005 or 0.5% probability that a value will be more than three standard deviations from the mean.

If we know the mean and standard deviation of a set of data and its distribution is normal, we can make some useful probability statements. For example, if the mean annual rainfall at a place is 50 cm per annum and the standard deviation is 15 cm we can say:
There is a 32% probability that next year's rainfall will be less than 35 cm or more than 65 cm.
There is a 5% probability that next year's rainfall will be less than 20 cm or more than 80 cm.
There is a 0.5% probability that next year's rainfall will be less than 5 cm or more than 95 cm.

Z-scores

A *z*-score (pronounced 'zee') for any specific value is the number of standard deviations that value is away from the mean. It may be found by calculating the formula:

$$z = \frac{x - \bar{x}}{\sigma}$$

where z = *z*-score
x = the specific value
$\bar{x}$ = the mean of all the values
σ = the standard deviation of all the values

z-scores are useful for telling us how *probable* events are. Two types of probability question are commonly asked:

1 What are the chances of an event occurring above or below a certain value? For example, we might like to know, for the purposes of flood prediction, the probability that a rainfall of over 100 cm will fall in London in any one year. Given the mean annual rainfall (62.5 cm) and the standard deviation (15 cm) the procedure is as follows:

Stage 1 Calculate the *z*-score for the specific value:

$$z = \frac{x - \bar{x}}{\sigma} = \frac{100 - 62.5}{15} = 2.5$$

This means that 100 cm is 2.5 standard deviations from the mean.

Stage 2 Look up the *z*-score in special *z*-score tables, column B (page 119). This tells us the probability that a value of more than 2.5 standard deviations above the mean could occur. The answer is 0.006 or 0.6%. We might, therefore, expect a rainfall of 100 cm or more less than once every 100 years (in fact, six times every 1000 years).

---

2 What are the chances of an event occurring *between* two values? For example, what is the chance of the traffic density at a junction being between 100 and 300 cars per hour at a certain time? The method is as follows:

Stage 1 Calculate the *z*-scores for each of the values, given $\bar{x} = 200$, $\sigma = 50$.

for $x = 100$, $$z = \frac{x - \bar{x}}{\sigma} = \frac{100 - 200}{50} = -2.0$$

for $x = 300$, $$z = \frac{x - \bar{x}}{\sigma} = \frac{300 - 200}{50} = +2.0$$

Stage 2 Look up the *z*-score in the *z*-score tables, column A, to find the probability of a value lying between the mean and the given value of *z*.

+2.0 = 0.477
−2.0 = 0.477

Stage 3 Add these probabilities to find the probability of a density occurring at or between these two values.

0.477
0.477+
————
0.954 or 95.4%

There is thus a 95.4% probability of the traffic density at the junction being between 100 and 300.

Probability paper

Probability paper, an example of which is given in figure 6.5, has several uses. By plotting data on it, it is possible to:

a See if a frequency distribution is approximately normal, and if it is –
b To read off its standard deviation and mean.
c To read off any required probabilities.

METHOD

Stage 1 Tabulate the data. The data given below is for annual rainfall at Bizerta, Tunisia.

Stage 2 Group the data into small, equal-sized classes. Record the number of observations within each class.

Stage 3 Add the number of observations within each class cumulatively. Convert these into percentages.

Stage 4 Plot these cumulative percentages on probability paper (fig. 6.5). Draw a straight line through the points, bisecting them, as close to the plotted points as possible.

Note the following:

a If the frequency distribution is approximately normal all the points plotted on the graph will be close to the straight line.
b The 50% line on the vertical axis cuts the straight line at the mean of the data (rainfall at Bizerta).
c The standard deviation is the difference between the mean and the value on the *x*-axis corresponding to either the 16% or 84% line on the vertical scale.
d Required probabilities can be read off from the probability paper. For example, the probability of rainfall being under 650 mm is just under 99% (read off the right-hand vertical scale) and of being over 650 mm is just over 1% (read off the left-hand scale).

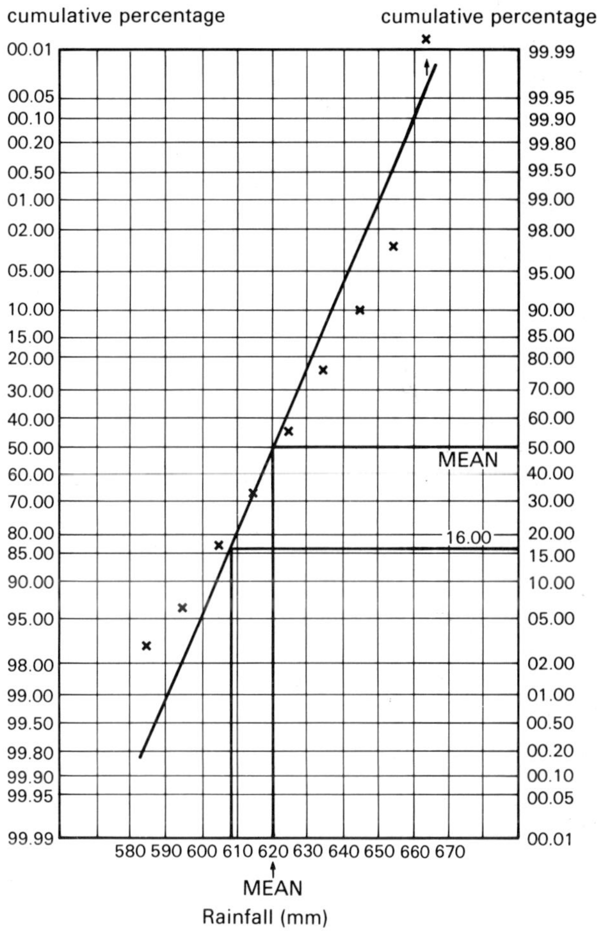

MEAN = 620 mm
Standard Deviation = 620 – 608
= 12 mm

Fig 6.5 Probability paper

EXAMPLE

Annual rainfall in millimetres, Bizerta, Tunisia, 1950–79

1950	630	1960	636	1970	584
1951	651	1961	628	1971	592
1952	601	1962	629	1972	615
1953	668	1963	608	1973	629
1954	648	1964	641	1974	614
1955	630	1965	629	1975	631
1956	638	1966	642	1976	617
1957	642	1967	630	1977	628
1958	604	1968	612	1978	620
1959	634	1969	653	1979	630

Grouped frequencies:

Rainfall groups	580–590	590–600	600–610	610–620	620–630	630–640	640–650	650–660	660–670
Frequency	1	1	3	5	7	6	4	2	1
Cumulative frequency	1	2	5	10	17	23	27	29	30
Cumulative percentage	3.3	6.7	16.7	33.3	56.7	76.7	90	96.7	100

6.5 Tests of significance

Very often in projects one collects two or more sets of sample data with the aim of comparing them and demonstrating contrasts. Examples might include:

a land values at the centre and outskirts of a town,
b temperatures in and out of a wood,
c pebble sizes at either end of a beach,
d crop yields on two different rock types.

Tests of significance are used to tell us whether the differences between two or more sets of sample data are truly significant or whether these differences *could* have occurred by chance. For example, if we sampled wheat yields on chalk and limestone the results might be as follows:

Rock	Yield (bushels/hectare)
Chalk	20
Limestone	18

Can we now say with confidence that the *actual* (rather than sampled) wheat yields are higher on chalk than on limestone? Or could it be that differences between the figures are due to *chance* and that another sample would give a different result? Tests of significance tell us the probability that differences between sample data are due to chance.

If we find that there *is* a significant probability that the relationship could have occurred by chance this can mean one of two things:

1 The relationship is not significant and there is little point in looking further for explanations of it.
2 Our sample is too small. If we took a larger sample we might well find that the result of the test of significance changes: the relationship becomes more certain.

It is not possible to tell which of these conclusions is the correct one from the results of the test itself. This is a good example of the way that statistics are only a tool and can never replace good geographical thinking.

The chi-squared (χ^2) test

The chi-squared test can only be used on data which has the following characteristics:

a The data must be in the form of frequencies counted in each of a number of categories. Data on the interval scale (data which has a precise numerical meaning, such as height above sea-level, the population size of a town or the temperature) can be grouped into categories to enable you to use this test.
b The total number of observations should be greater than 20.
c The expected frequency in any one fraction (described in stage 3 below) should not be less than 5.
d The observations should not be such that one influences another.

χ^2 for two variables

METHOD

Stage 1 Write out the hypothesis you are hoping to prove, i.e. the fact that there is a significant difference between the sample data sets.

Stage 2 Tabulate the data as shown in the example. The data being tested for significance is called the 'observed frequency', and the column is thus headed 'O'.

Stage 3 Calculate the number of counts you would *expect* to find in each category if the categories had no impact on these. This is the 'expected frequency' ('E').

Stage 4 Calculate the formula:

$$\chi^2 = \Sigma\frac{(O - E)^2}{E}$$

where χ^2 = chi-squared figure
Σ = the sum of
O = observed frequency
E = expected frequency

Stage 5 Calculate what is called the 'degrees of freedom'. This is simply one less than the total number of categories:

$$df = n - 1$$

where df = degrees of freedom
n = no. of categories in the test

Stage 6 Turn to appendix 4 and, using the calculated value of χ^2 and the degrees of freedom, read off the probability that the data frequencies you are testing could have occurred by chance.

EXAMPLE

You have visited four equal-sized areas, each on a different rock type. You counted the number of streams on each area. The results were as follows:

Rock type	**No. of streams**
Chalk	7
Granite	58
Limestone	15
Sandstone	20

You now wish to know if these results are a true reflection of the nature of each rock type or whether they could simply be the result of chance.

The χ^2 test can be used:

a The data is in the form of counts.
b The total number of streams observed exceeds 20.
c The expected frequency in any one fraction exceeds 4. This is the number you would *expect* if rock type had no influence on stream densities. In this case it is the total number of streams (100) divided by the number of rock types (4).
d The observations are independent. (The number of streams on one rock type does not influence the number of streams on another.)

Stage 1 The hypothesis we are trying to prove is that there is a significant difference between the sample data sets, i.e. rock type does influence stream density.

Stage 2 Tabulation: see stage 4.

Stage 3 If rock type had no influence on stream density one would expect an equal number of streams on each rock. There are 100 streams. The areas examined on each rock type are the same. You would therefore expect 25 streams on each rock type, E = 25. If the areas surveyed on each rock type were not equal the expected frequencies would have to be divided amongst the rock types proportional to their areas.

Stage 4

Rock type	**Observed frequency (O)**	**Expected frequency (E)**	$\frac{(O - E)^2}{E}$
Chalk	7	25	$(7 - 25)^2 \div 25 = 13.0$
Granite	58	25	$(58 - 25)^2 \div 25 = 43.6$
Limestone	15	25	$(15 - 25)^2 \div 25 = 4.0$
Sandstone	20	25	$(20 - 25)^2 \div 25 = 1.0$
			Σ61.6

$$\chi^2 = \Sigma\frac{(O - E)^2}{E} = 61.6$$

Stage 5 df $= n - 1$
$= 4 - 1$
$= 3$

Stage 6 From the graph in appendix 4 read off the degrees of freedom (3) on the horizontal axis against the χ^2 value (61.6) on the vertical axis. The resulting point is above the line marked 0.1 chance in 100. This means that the probability that the data given above could be due to chance alone is less than 1 in 1000.

χ^2 for more than two variables

In the previous example only two variables were used (rock type and stream density). In this example we use three (temperature groupings and readings taken at two places, one near to and the other away from a lake).

METHOD

Stage 1 Write out the hypothesis you are trying to prove.

Stage 2 Tabulate the data as shown in the example.

Stage 3 Calculate the expected frequency (E). This is found by multiplying the sum of the row (Σr) by the sum of the column (Σk) in which the observed frequency lies and dividing by N, the sum of all observed frequencies.

Stage 4 Calculate the formula:

$$\chi^2 = r\Sigma k\Sigma\left(\frac{(O - E)^2}{E}\right)$$

where $r\Sigma k\Sigma$ = the sum of the fraction for all values of r and k
O = observed frequency
E = expected frequency

Stage 5 Calculate the degrees of freedom (df). This is the number of rows less one, times the number of columns less one:

$$\text{df} = (r - 1)(k - 1)$$

Stage 6 Turn to appendix 4 and, using the calculated value of χ^2 and the degrees of freedom, read off the probability that the data frequencies you are testing could have occurred by chance.

EXAMPLE

The aim of this project was to discover whether a lake had any impact on the temperature of the surrounding land. Temperature readings were taken at two points, one on the lake shore, the other 500 metres away. Thirty readings were taken at each place, spaced over a period of five days. The results were grouped into 0.5°C intervals and tabulated as follows:

Temperature °C	**Near lake**	**Away from lake**
13.0–13.4	4	2
13.5–13.9	3	8
14.0–14.4	4	7
14.5–14.9	12	7
15.0–15.4	7	6

Stage 1 Hypothesis: that there is a significant difference between the temperatures near to and away from the lake.

Stage 2 Tabulation of data:

		Near lake		Away from lake		
Row number (r)	Column number (k)	O k_1	E	O k_2	E	Σr
r_1	13.0–13.4	4	3	2	3	6
r_2	13.5–13.9	3	5.5	8	5.5	11
r_3	14.0–14.4	4	5.5	7	5.5	11
r_4	14.5–14.9	12	9.5	7	9.5	19
r_5	15.0–15.4	7	6.5	6	6.5	13
		$\Sigma k_1 = 30$		$\Sigma k_2 = 30$		$N = 60$

Stage 3 For example, the observed frequency for the value k_1, r_1 is 4. The expected frequency is the sum of its row (6) multiplied by the sum of its column (30), divided by the sum of all observed frequencies (60):

$$E = \frac{30 \times 6}{60} = \frac{180}{60} = 3$$

Stage 4 $\chi^2 = r \Sigma k \Sigma \left(\frac{(O - E)^2}{E}\right)$

$$= \frac{(4-3)^2}{3} + \frac{(2-3)^2}{3} + \frac{(3-5.5)^2}{5.5} + \frac{(8-5.5)^2}{5.5} + \frac{(4-5.5)^2}{5.5} + \frac{(7-5.5)^2}{5.5} + \frac{(12-9.5)^2}{9.5} + \frac{(7-9.5)^2}{9.5} + \frac{(7-6.5)^2}{6.5} + \frac{(6-6.5)^2}{6.5}$$

$$= 5.15$$

Stage 5 $df = (r-1)(k-1)$
$= (5-1)(2-1)$
$= 4$

Stage 6 Read off the graph in appendix 4 the degrees of freedom (4) on the horizontal axis against the χ^2 value (5.15) on the vertical axis. The resulting point is below the line marked 10 chances in 100. This means that the probability that the apparent difference between the two sets of temperature readings could just be due to chance is high: in fact, a more than one in ten chance.

6.6 Correlation

Two things *correlate* when they vary together, such as temperature decreasing with altitude or land values falling with distance from the city centre. If, as one variable increases in value so does the other, this is a *positive* correlation. If one goes up as the other goes down, this is a *negative* correlation. Some things correlate fairly exactly: temperature falling with altitude, for example. Other things correlate but not very well, such as people's age and height.

These correlations can be seen if the two variables are plotted on a graph (fig. 6.6). The level of correlation can also be expressed as a numerical index and this is what the following technique does: it expresses a relationship as a number, known as a 'correlation coefficient'. This is useful for three reasons:

a It is more precise than a graph: while two graphs showing correlations may look similar, the correlation coefficients for the sets of data may well be slightly different.

b If we wanted to compare several pairs of data, such as the relationship between temperature and altitude on twenty slopes, it would be far easier to compare twenty *numbers* expressing these relationships than twenty *graphs*.

c It is possible to test the correlation to see if it is really significant or whether it could have occurred by chance alone.

A WARNING

The fact that two things correlate proves nothing. It can never be concluded from statistical evidence alone that because two things correlate one must be affecting the other. If we found that crop yields decreased with height up a mountain we would not know without further research whether this was due to the fall in temperature, or steeper slopes, or changes in precipitation. All projects involving correlations *must* be supplemented by research which seeks to uncover the processes behind the correlation.

Spearman's rank correlation coefficient (r_s)

This technique is among the most reliable methods of calculating a correlation coefficient, one number which will summarise the strength and direction of any correlation between two variables.

METHOD

Stage 1 Tabulate the data as shown in the example. Rank the two data sets independently, giving the highest value a rank of 1 and so on.

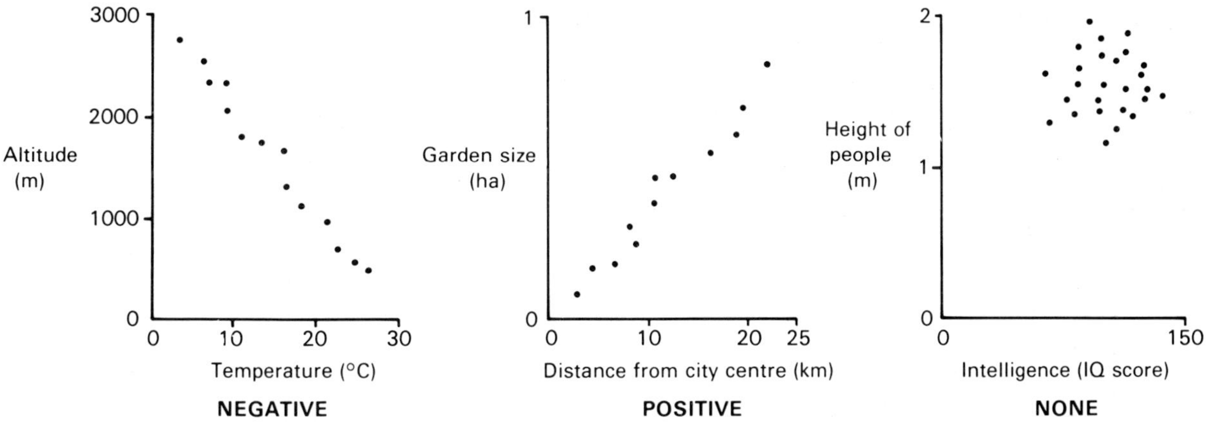

Fig 6.6 Types of correlation

Stage 2 Find the difference between the ranks of each of the paired variables (d). Square these differences (d^2) and sum them (Σd^2).

Stage 3 Calculate the coefficient (r_s) from the formula:

$$r = 1 - \frac{6 \Sigma d^2}{n^3 - n}$$

where d = the difference in rank of the values of each matched pair

n = the number of pairs

Σ = the sum of

The result can be interpreted from the scale:

+1.0 →	0 →	−1.0
Perfect positive correlation	No correlation	Perfect negative correlation

It is now desirable to determine whether the correlation you have calculated is really significant, or whether it could have occurred by chance.

Stage 4 Decide on the rejection level (α). This is simply how certain you wish to be that the correlation you have calculated could not just have occurred by chance. Thus, if you wish to be 95% certain your rejection level is calculated as follows:

$$\alpha = \frac{100 - 95}{100} = 0.05$$

Stage 5 Calculate the formula for t:

$$t = r_s \sqrt{\left(\frac{n - 2}{1 - r_s^2}\right)}$$

where r_s = Spearman's rank correlation coefficient

n = number of pairs

Stage 6 Calculate the degrees of freedom (df):

$$\text{df} = n - 2$$

where n = the number of pairs

Stage 7 Look up the critical value in the t-tables (page 123), using the degrees of freedom (df, stage 6) and rejection level (α, stage 4). If your t-value was positive then the correlation is significant (at the chosen level) if the critical value is less than the t-value. If your t-value was negative then the correlation is significant if the t-value is less than the negative equivalent of the critical value in the tables. If the correlation is not significant this may mean one of two things:

a The relationship is not a good one and it is thus not really worth pursuing it any further.

b The size of sample you are using is too small to permit you to prove a correlation. If you increase the sample size a statistically significant correlation may then emerge.

It is not possible to tell which of these conclusions is the correct one from the technique alone. It requires intelligent geographical thinking on your part to decide this.

EXAMPLE

Data: Population size and number of services in each of 12 settlements

Population	No. of services
350	3
5 632	41
6 793	43
10 714	87
220	4
15 739	114
8 763	72
7 982	81
6 781	73
4 981	35
1 016	11
2 362	19

Stages 1–2

1	2	3	4	5	6
Settlement population	Rank	No. of services	Rank	Difference between ranks(d)	d^2
220	12	4	11	1	1
350	11	3	12	1	1
1 016	10	11	10	0	0
2 362	9	19	9	0	0
4 981	8	35	8	0	0
5 632	7	41	7	0	0
6 781	6	73	4	2	4
6 793	5	43	6	1	1
7 982	4	81	3	1	1
8 763	3	72	5	2	4
10 714	2	87	2	0	0
15 739	1	114	1	0	0
					$\Sigma d^2 = 12$

Stage 3

$$r_s = 1 - \frac{6\,\Sigma\,d^2}{n^3 - n}$$

$$= 1 - \frac{6 \times 12}{12^3 - 12}$$

$= +0.96$ (a strong positive correlation)

Stage 4 Rejection level (α) = 95%
= 0.05

Stage 5

$$t = r_s\sqrt{\left(\frac{n-2}{1-r_s^2}\right)}$$

$$= 0.96\sqrt{\left(\frac{12-2}{1-0.96^2}\right)}$$

$= 10.73$

Stage 6

$$\text{df} = (n - 2)$$
$$= (12 - 2)$$
$$= 10$$

Stage 7

df = 10
rejection level = 0.05
∴ critical value of t = 2.23

The critical value is less than our t-value (10.73). We can therefore conclude that there is a significant correlation between settlement size and the number of services offered in each.

6.7 Regression

Correlation analysis tells us if two variables are significantly related. Regression analysis goes one step further and:

a tells us how to predict a value of one of these variables from the corresponding value of the other.

b points out deviations (called 'residuals') from the average relationship between the two variables.

Regression analysis is only relevant to data sets which correlate. If two such sets are so related they tend towards a diagonal line pattern if plotted on a graph (fig. 6.7). A perfect positive correlation is plotted in A and a perfect negative correlation in B. A less perfect but still generally positive correlation is shown in C. The lines drawn on these graphs are 'regression lines'.

A regression line is a straight 'best-fit' line, positioned in such a way that the distance between it and all the points on the graph is at a minimum. The distances of the points from the regression line, measured on the vertical scale, are 'residual values'. They show how much the points deviate from the best-fit line.

Remember that if you are plotting two variables, changes in one of which seem to cause changes in the other, the variable *causing* the change is called the independent variable and is plotted on the horizontal or 'x'-axis, and the one which is *caused* to change is called the dependent variable and is plotted on the vertical or 'y'-axis.

The best-fit regression line can be drawn in by eye, but this is relatively crude. There is a more accurate system called the 'semi-averages method'. This method is described here and a worked example is given in figure 6.8.

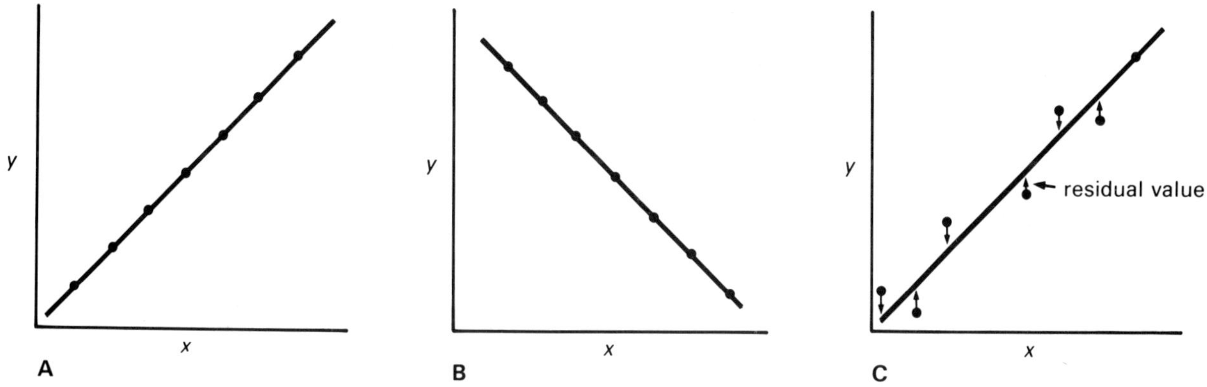

Fig 6.7 Graphs showing different sorts of correlating relationships and regression lines

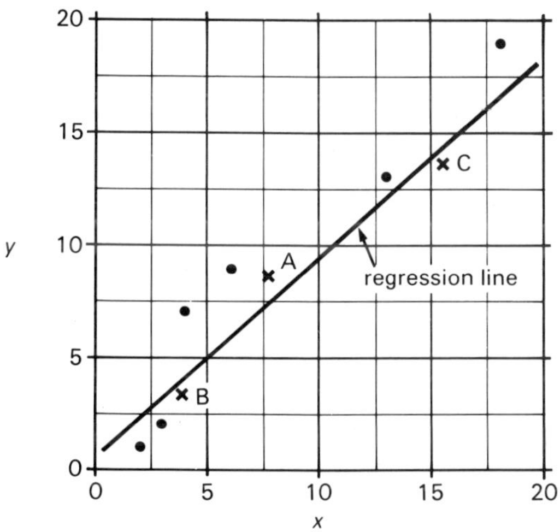

Fig 6.8 The semi-averages method for drawing a regression line

METHOD

Stage 1 Plot the data on a graph.

Stage 2 Find the means of the two sets of data ($\bar{x}$ and $\bar{y}$). Use these as coordinates to plot a mean point (A) on the graph.

Stage 3 Find the lower semi-average. This is done by finding the mean of all the values of x below $\bar{x}$ and the mean of all the values of y below $\bar{y}$. These figures are then used as coordinates to plot the lower semi-average point (B).

Stage 4 Find the upper semi-average. This is done by finding the mean of all the values of x above $\bar{x}$ and the mean of all the values of y above $\bar{y}$. These figures are then used as coordinates to plot the upper semi-average point (C).

Stage 5 These three points (A, B, C) form an almost straight line. Draw a best-fit line by eye through these. This is the regression line.

Now, given any value of x we can read off the roughly corresponding value of y (and vice versa).

EXAMPLE

Data:

Unemployment (%)	x	2	3	4	6	13	18
Annual outbreaks of street rioting	y	1	2	7	9	13	19

Stage 1 See figure 6.8

Stage 2 Mean of x values ($\bar{x}$) = 7.7, Mean of y values ($\bar{y}$) = 8.5 } coordinates for point A

Stage 3 Mean of all the values of x below $\bar{x}$ = 3.8, Mean of all the values of y below $\bar{y}$ = 3.3 } coordinates for point B

Stage 4 Mean of all the values of x above $\bar{x}$ = 15.5, Mean of all the values of y above $\bar{y}$ = 13.7 } coordinates for point C

Stage 5 See graph.

This enables us to *predict* relationships without actually measuring them – the relationship between city size and level of crime, or between stream discharge and amounts of load, for example. It may, however, be dangerous to extrapolate too far. If, in our graph, the x-axis represents levels of unemployment and the y-axis the number of street riots per annum, we could not fairly go on to say what the level of rioting would be if unemployment reached 50%.

Deviation from the regression line (residuals) can also be observed. These might be places which have high unemployment but very little rioting, or places with low unemployment which are nevertheless particularly volatile. Often these form more worthwhile points of discussion than the trend shown by the best-fit line on the graph, which may simply be what we expected anyway.

6.8 Exercises

1 In order to analyse the way that longshore drift selectively carries beach material of a given size, the long axes of 10 pebbles were measured at each end of a beach. Calculate the mean, mode, median, inter-quartile range and standard deviation of the two data sets (pages 91–3).

Pebble long axis (cm)

Western end of beach	Eastern end of beach
0.2	3.4
1.3	0.7
2.1	6.1
0.7	10.8
4.1	3.4
2.3	7.3
3.8	12.7
0.9	2.9
6.2	15.8
5.9	6.7

2 It is thought that the amount a stream meanders is connected to the nature of its load. A survey was conducted which recorded the amount of sand in the load of streams and a measure of meandering, the 'sinuosity index'.
The results were as follows:

Sample no.	1	2	3	4	5	6	7	8	9	10	11	12
Sinuosity index	120	115	130	124	138	160	155	148	150	162	142	137
Percentage sand in sample	10	5	9	8	12	19	17	16	19	18	7	6

Use Spearman's rank correlation coefficient (page 100) to calculate the degree of association between the proportion of sand in the samples and the sinuosity index.

3 When planning an agricultural development project in highland Kenya it was found that the mean daily temperature was 20°C and the standard deviation 10°C. What is the probability (page 95) of the daily temperature falling below 0°C (a critical temperature, causing crop damage)?

4 Construct a regression line for the following data using the semi-averages method (page 103):

Distance from village (km)	1	2	3	4	5	6	7
No. of marriage partners marrying people in the village in a five-year period	10	8	3	4	2	1	1

5 A study was made of the number of antique shops in five equal-sized towns in Yorkshire to discover whether differences existed between the locations. The information obtained was as follows:

Town A	21
Town B	15
Town C	17
Town D	8
Town E	25

There is an apparent difference in terms of numbers of antique shops. Use a chi-squared test (page 98) to determine whether this is significant or not.

6 A survey of rateable values in London seemed to suggest that they declined with distance from the centre. The data obtained was as follows:

Distance from centre (km)	Rateable values (£)
0–1	4000
1–2	1000
2–3	500
3–4	600
4–5	300
5–6	400
6–7	100

Calculate the Spearman's rank correlation coefficient (page 100) for this data and check its significance.

7 Spatial analysis

If we were interested in examining the distribution of phenomena such as settlements, shops or landforms, the nature of a road network or flows of traffic along a route, we could use spatial analysis. Spatial analysis enables us to give *quantitative* (numerical) measures to these things and thus adds a level of precision which simple verbal description cannot give. It may also help us to discover something about the distribution which was not immediately obvious.

7.1 Analysis of point distributions

The following methods of analysis are applicable to any geographical feature which forms a point distribution (rather than a line or area), such as towns, factories, shops, drumlins or erratics.

1 Central point

This method locates the central point of a distribution. It is most useful when comparing the distribution of two or more things, for it enables us to see the patterns more clearly. The simplest method is to find the 'median centre':

METHOD

Stage 1 Plot the points on a map.

Stage 2 Draw two lines, running north–south and east–west respectively, such that they bisect the point distribution, leaving an equal number of points on either side of each line. If the total number of points is an odd number the lines must pass through a point. If the total is an even number the lines must be halfway between two points. Where the two lines intersect is the median centre (fig. 7.1).

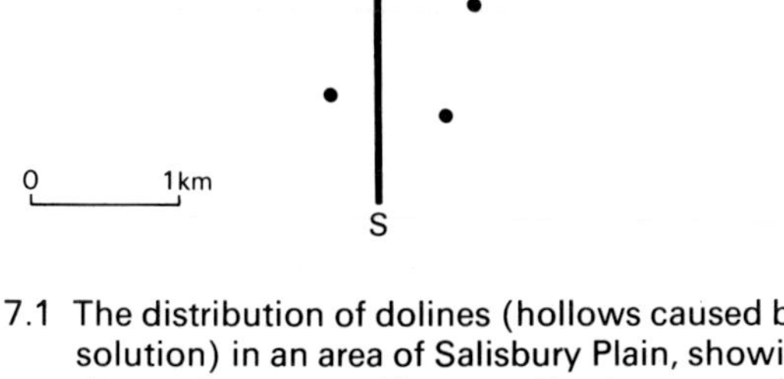

Fig 7.1 The distribution of dolines (hollows caused by solution) in an area of Salisbury Plain, showing the median centre. There are 5 points on each side of the N–S line and 5 points on each side of the E–W line.

2 Dispersion of points relative to each other: the nearest neighbour index

Figure 7.7 shows the distributions of villages on a plain. In A the villages are obviously well spaced, but in a fairly random manner. In B they appear, with two exceptions, to be fairly clustered. If one were comparing a number of distributions of this type it might be desirable to express each pattern in a single index. The nearest neighbour index does this, for it tells us in one statistic how clustered, uniform or random in distribution a series of points is.

Fig 7.2 Services such as banks tend to cluster together. Their distribution can be analysed using nearest neighbour analysis.

The nearest neighbour index has been calculated for the pattern in figure 7.7 A. The process is as follows:

METHOD

Stage 1 Plot the points on a map and number them.

Stage 2 Draw up a table as shown and find the distance from each point to its nearest neighbouring point. It is quite possible that a point will be the nearest neighbour of several other points, or that two points will be each other's nearest neighbour.

Stage 3 Sum these nearest neighbour distances (ΣD) and divide the result by the number of points (n). This gives you the mean observed distance of all points to their nearest neighbours ($\bar{D}$).

Stage 4 Calculate the nearest neighbour index from the following formula:

$$\text{NNI} = 2\,\bar{D}\sqrt{\frac{n}{A}}$$

where NNI = the nearest neighbour index
$\bar{D}$ = the mean observed nearest neighbour distance
$\sqrt{\ }$ = square root of
n = the total number of points
A = the area of the map on which the points lie

The units used for this must be the same as those used to calculate distances in stage 2 above.

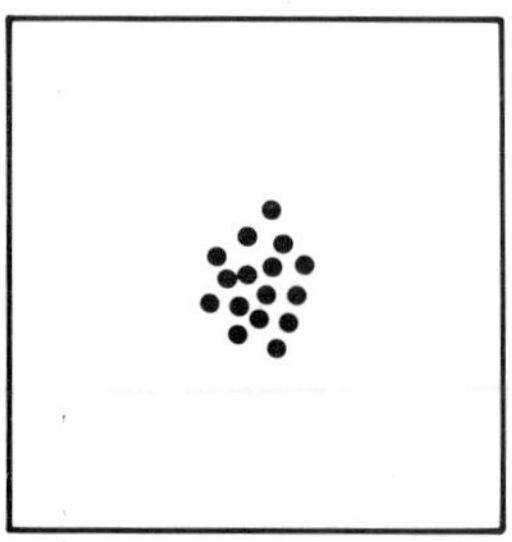

0

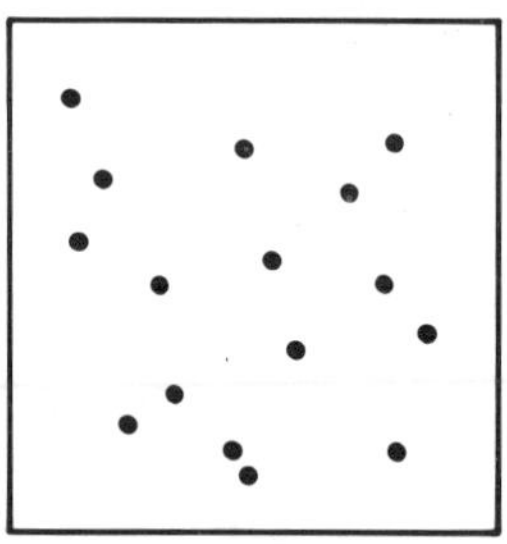

1.0

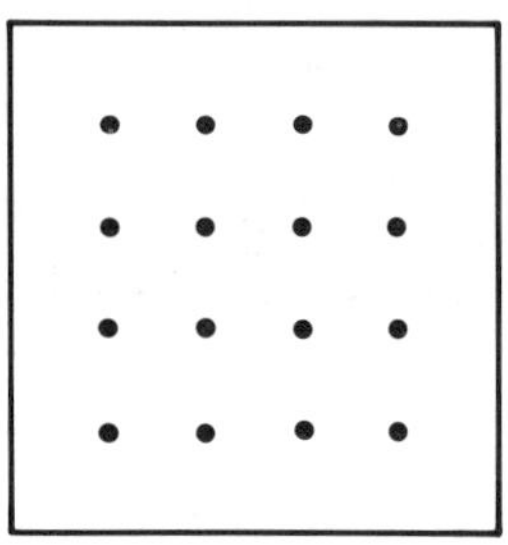

2.15

Fig 7.3 The scale of nearest neighbour index values

The result can be interpreted from the following scale:

Nearest neighbour index (refer to figure 7.3)

0	=	points completely clustered together
1.0	=	points in a completely random distribution
↓ 2.15	=	points in a completely uniform distribution

Thus if the NNI was 0.7 the verbal interpretation would be 'more nearly random than clustered'. If it was 1.9 it would be 'more nearly uniform than random', and so on.

There are a number of problems with the nearest neighbour index and it is as well to be aware of these:

a It cannot distinguish between a single and multi-clustered pattern. Both the following distributions, although different, have a NNI of approximately 0 (fig. 7.4).

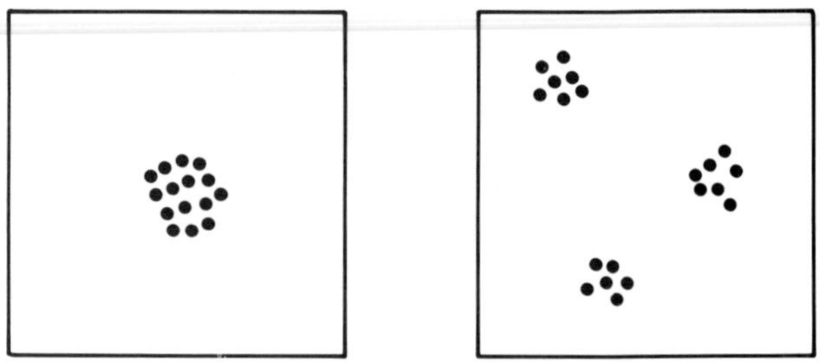

Fig 7.4 Single and multi-clustered point patterns

b An index of 1.0 does not *always* necessarily mean that the distribution is totally random. Two sub-patterns on the map, when combined in one index, may give a false impression of randomness. Thus figure 7.5 produces a NNI of 1.0 although it is clearly not random.

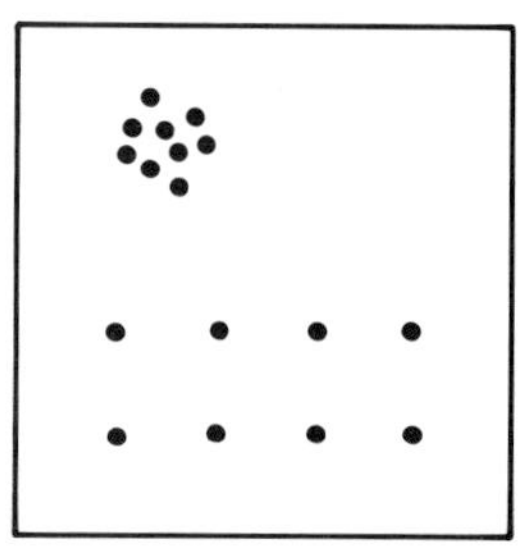

Fig 7.5 Two types of point pattern on one map

c Neither should a distribution with a NNI of 1.0 be interpreted as being *caused by* totally random or chance factors. It is possible for a settlement distribution to have a NNI of 1.0 and the settlements to be closely related to another factor such as the emergence of springs. The settlement location cannot be said to be due to random factors. It is the spring location that is more fundamentally random.

d The NNI obtained depends very much on the area of the map in which the dots lie. Figure 7.6 shows the same distribution but on different scale maps. The small scale map gives an impression of *clustering*, while the large scale map suggests the same points have a *random* distribution.

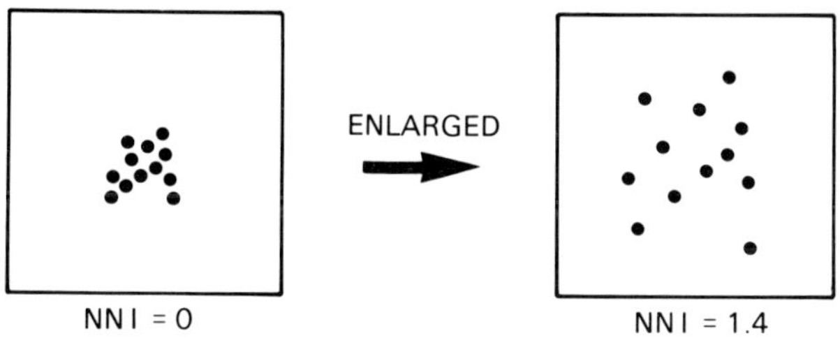

Fig 7.6 A point pattern shown at two different scales

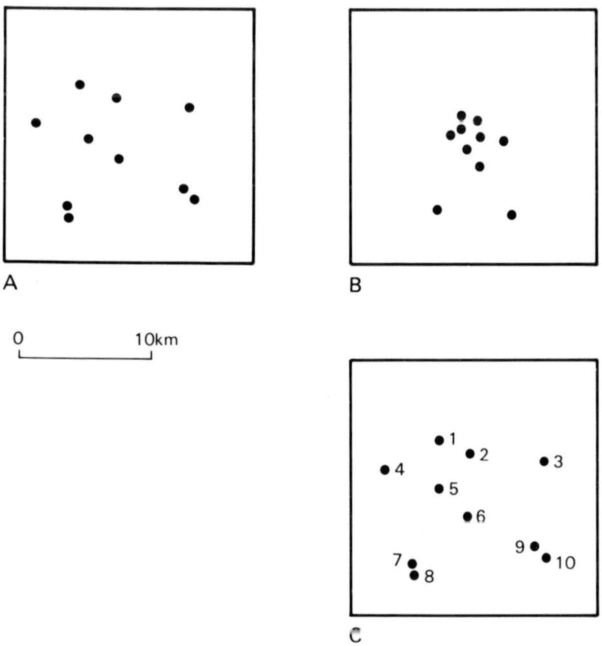

Fig 7.7 Calculation of the nearest neighbour index

EXAMPLE (refer to figure 7.7 C)

Point	Nearest neighbour	Distance apart, km
1	2	2.5
2	1	2.5
3	2	5.0
4	5	4.0
5	6	3.0
6	5	3.0
7	8	1.0
8	7	1.0
9	10	1.0
10	9	1.0
		24.0

$$\Sigma D = 24.0$$
$$n = 10$$
$$\bar{D} = \frac{24}{10}$$
$$= 2.4 \text{ km}$$
$$A = 256 \text{ km}^2$$
$$\text{NNI} = 2\bar{D}\sqrt{\frac{n}{A}}$$
$$= 2 \times 2.4\sqrt{\frac{10}{256}}$$
$$= 2 \times 2.4 \times 0.2$$
$$= 0.96$$
= an almost random distribution

7.2 The index of dissimilarity

The index of dissimilarity is used to compare the distribution of two variables, notably different social or ethnic groups.

METHOD

Stage 1 The data must be tabulated in the form shown in the example. In other words, it must be in percentage form and the total for each variable must be 100% (the whole distribution is accounted for). Call one variable x, the other y.

Stage 2 Find the difference between the paired variables and add these differences. Halve the result. This gives the index of dissimilarity. The formula is:

$$\text{Id} = \tfrac{1}{2}\Sigma|x_i - y_i|$$

where Id = the index of dissimilarity
Σ = the sum of
x_i, y_i = the paired variables
$|\ |$ = irrespective of sign (i.e. always take the smaller of the two figures away from the larger)

Stage 3 The index of dissimilarity varies between 0 and 100%. It represents the proportion of the population group x which would have to change its location in order to replicate (copy) the distribution of the group with which it is being compared (y).

EXAMPLE

Calculation of the index of dissimilarity (USA 1980)

Region	x % of black pop.	y % of white pop.	|x − y|
South	53	28	25
North East	19	25	6
North Central	20	29	9
West	8	18	10
	100	100	50

$$\Sigma|x_i - y_i| = 50$$
$$Id = \tfrac{1}{2}\Sigma|x_i - y_i|$$
$$Id = 50 \div 2$$
$$= 25$$

This result means that 25% of the black population of the United States would have to change the region in which they live in order to have the same relative distribution as the white population, or vice versa.

MERITS:

a The index is simple to calculate.
b It has a direct verbal meaning (stage 3 above).
c The result is not affected by the relative sizes of the two groups that are being compared.

LIMITATION: The result is affected by the size of the areal units within which the data was collected. The larger these units, the smaller the resulting index.

7.3 The location quotient

The location quotient (LQ) is a measure of how concentrated a particular phenomenon is in one area, for example, how concentrated the printing industry of London is in Fleet Street or how concentrated coal mining in northern England is in County Durham. To calculate the location quotient one needs to know the distribution of the phenomenon under study (the printing industry) *and* the distribution of the wider population of which the phenomenon is just a part (all industry).

Figure 7.8 shows four enumeration districts which cover the whole of a town in NW England containing a large Italian population. To find the location quotient for Italians in each enumeration district we need to know the proportion of Italians in each and the proportion of the *total* population in each. We then divide the former by the latter.

In theory the location quotient ranges between zero and infinity. If an area has a LQ greater than 1.0 then the sub-group of the total population is over-represented. The larger the LQ the greater the concentration of the sub-group. In figure 7.8 districts 1 and 2 are over-represented in terms of the proportion of the population born in Italy while districts 3 and 4 are under-represented.

EXAMPLE

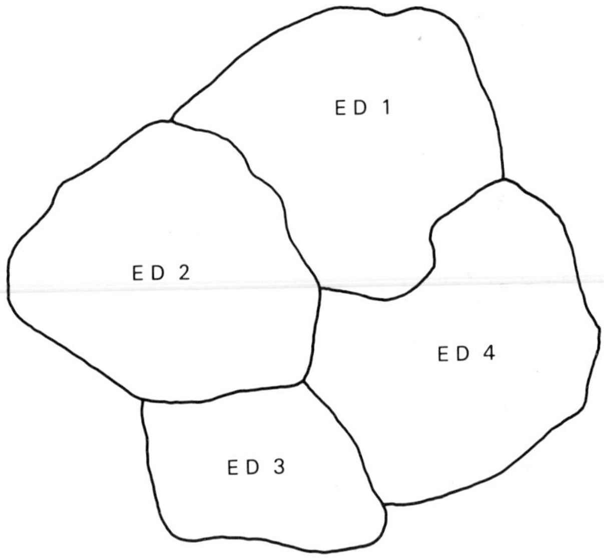

Enumeration district (ED)	% of Italians in each ED	% of total population each ED	Location quotient (LQ)
1	35	25	(35 ÷ 25) = 1.4
2	40	25	(40 ÷ 25) = 1.6
3	20	25	(20 ÷ 25) = 0.8
4	5	25	(5 ÷ 25) = 0.2

Fig 7.8 Calculation of the location quotient for Italians in four enumeration districts in a town in NW England

7.4 Gravity models

Gravity models help predict the amount of movement or 'interaction' between two places. They can tell you, for example, what proportion of people from a village might shop in a nearby town X rather than another town Y, or how much traffic we would expect between towns A and B relative to that between towns A and C.

Gravity models are based on the law of physics that the attraction between two bodies (such as planets) is proportional to the product of their masses divided by the square of the distance between them. In equation form:

$$\text{Attraction between A and B} = \frac{\text{mass of A} \times \text{mass of B}}{(\text{distance from A to B})^2}$$

This may be rewritten for two settlements as follows:

$$\text{Interaction between towns A and B} = \frac{\text{population of A} \times \text{population of B}}{(\text{distance from A to B})^2}$$

EXAMPLE

Figure 7.9 shows three settlements and their population figures. The interaction between A, B and C is as follows:

$$I\,A,B = \frac{pApB}{(dAB)^2} = \frac{1000 \times 500}{(2.8)^2} = 63\,776$$

$$I\,A,C = \frac{pApC}{(dAC)^2} = \frac{1000 \times 100}{(2.8)^2} = 12\,755$$

where I = interaction
p = population
d = distance

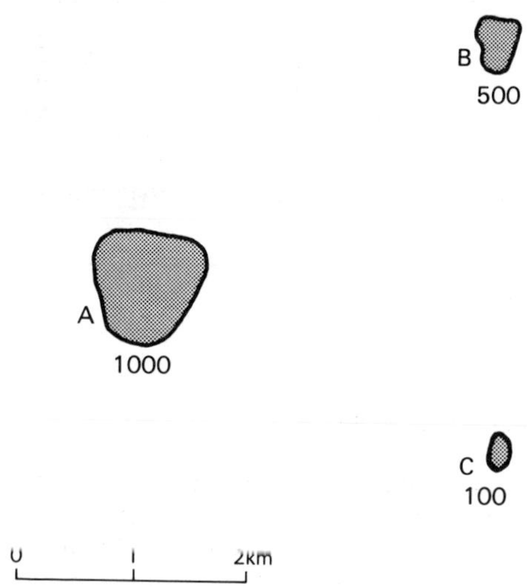

Fig 7.9 Three settlements with population sizes given

The resultant figures do not mean anything in themselves. All we can say from this is that the hypothetical interaction between A and B (telephone calls, shopping movements, etc.) is about five times that between A and C. This is because the population of B is larger than that of C.

The model is clearly based on two assumptions:

a Bigger places interact more with each other than smaller places.
b Nearer places interact more with each other than more distant places.

Project suggestion

You could test the gravity model in a project to see whether these assumptions are true in reality. To do this you will have to measure the interaction between the two places, possibly analysing such things as bus timetables (to find out the number and frequency of services between the two), or doing a survey of the residences of shoppers in the two towns. You may find that the model is more accurate if, instead of straight-line distance between the two places, you use travel time or cost instead.

7.5 Transport network analysis

Connectivity and accessibility

Topological networks: terminology For the accurate analysis and planning of transport networks a series of simple techniques has been devised. Connected with all these techniques are a number of important terms and concepts:

1 **Topology** Topology is the study of the relative positions of places. For the purpose of analysis, transport networks are commonly subject to a 'topological transformation'. In such a transformation *distance* and *direction* are disregarded, but the *relative position* of a place is retained (fig. 7.11). The London Underground map is topologically transformed (fig. 7.10).

   The topologically transformed map is known as a 'graph' and the techniques used to analyse such a map are part of 'graph theory'.

2 Settlements on a network are known as 'nodes' or 'vertices'. Usually route junctions and route endings are also counted as nodes (fig. 7.11).

3 The stretch of route between two nodes is known as an 'edge', 'arc' or 'link'. For the purpose of analysis edges which exactly duplicate each other by linking the same two nodes are usually counted as one edge (fig. 7.12).

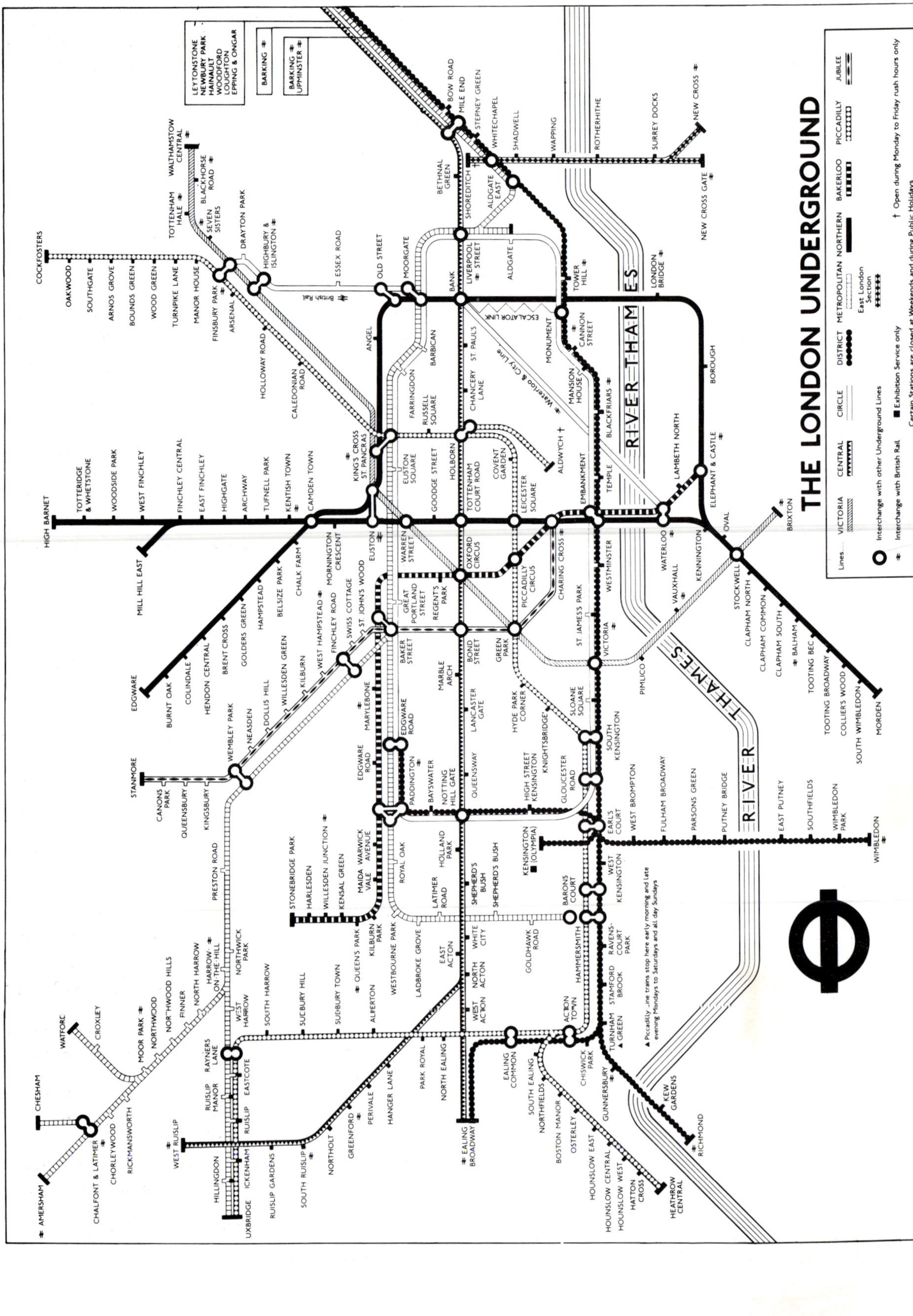

Fig 7.10 A topological transformation of the London Underground network

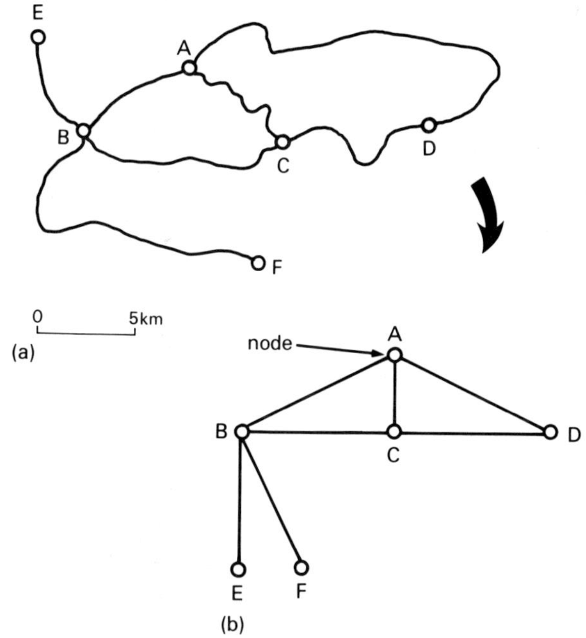

Fig 7.11 A real route network (a), topologically transformed (b)

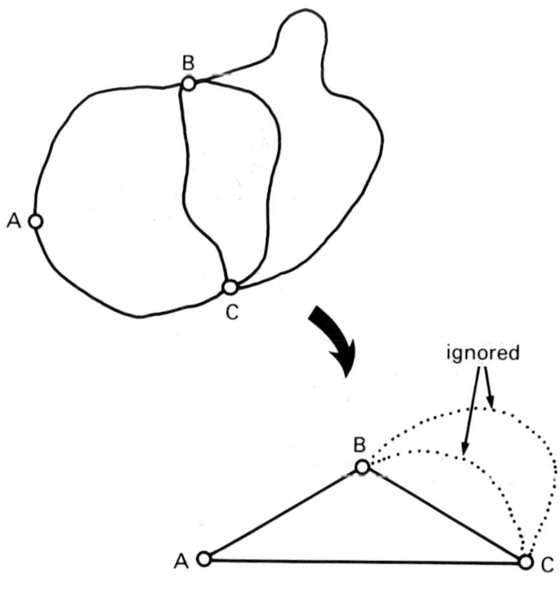

Fig 7.12 A topological transformation illustrating that duplicate edges are ignored

4 There are two basic types of graph:
 a Planar: a two-dimensional network with all routes actually joining up in real space, such as a road system with no flyovers or underpasses (fig. 7.13).

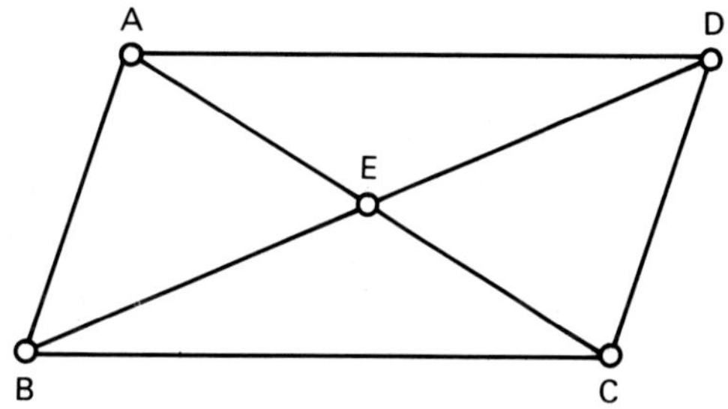

Fig 7.13 A planar graph

 b Non-planar: a three-dimensional network with one or more routes which cross over but do not actually join up with other routes in space, such as an airway network or a road flyover. In figure 7.14 edge A–C does not actually link up with edge B–D, which passes over A–C as a semi-circle.

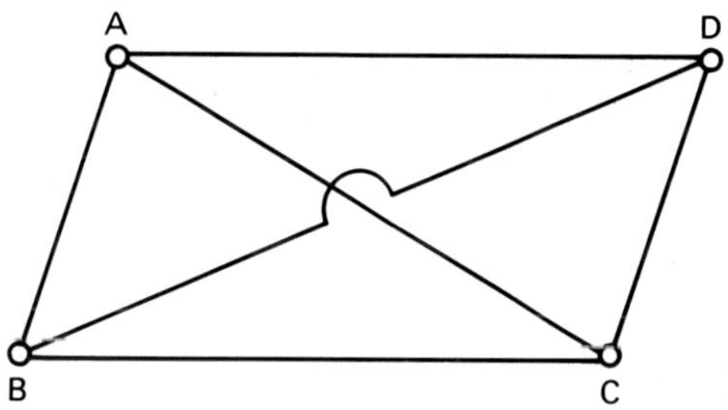

Fig 7.14 A non-planar graph

5 Finally, a network in which the *direction* of traffic flow along the edges is *specified* is a 'directed graph'. This would be true, for example, of a one-way street system (fig. 7.15).

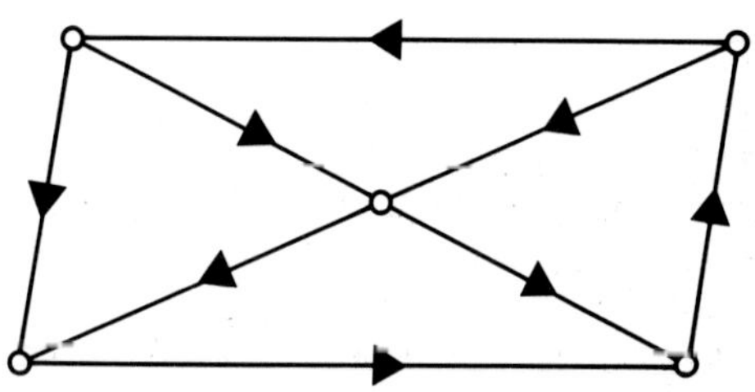

Fig 7.15 A directed graph

Analysis of topological networks A number of quantitative techniques have been developed to analyse these networks. Each one will be illustrated with reference to the simple graph shown in figure 7.16.

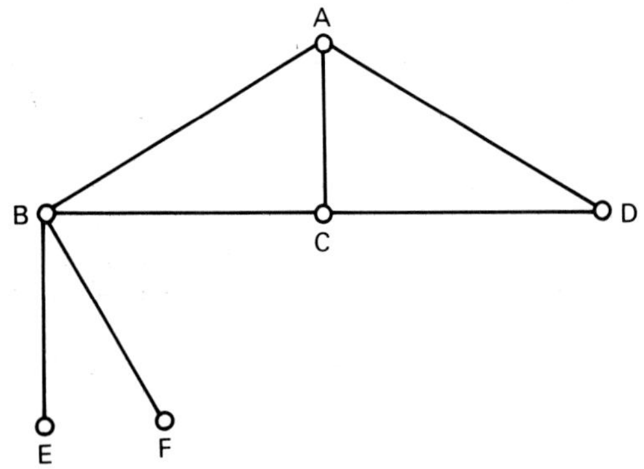

Fig 7.16 A simple network

Measuring connectivity The following techniques all measure how interconnected the network is. Maximum connectivity is found when all nodes are connected directly to all other nodes; minimum connectivity when no nodes are connected to any other nodes.

1 Gamma index (γ): There are two formulae for the gamma index, one for planar and the other for non-planar graphs:

$$\text{Planar graphs:} \quad \gamma = \frac{e}{\frac{1}{2}(n^2 - n)}$$

$$\text{Non-planar graphs:} \quad \gamma = \frac{e}{3(n - 2)}$$

where e = the number of edges
n = the number of nodes

Our example is a planar graph. The γ index is therefore as follows:

$$\gamma = \frac{7}{\frac{1}{2}(36 - 6)} = \frac{7}{15} = 0.47$$

The index ranges from a minimum connectivity of 0.0 to a maximum connectivity of 1.0. This index is often multiplied by 100, converting it to a percentage figure with a range from 0.0% to 100%. Thus our example has a connectivity of 47%.

2 Diameter (d): The diameter of a network is the number of edges one has to pass along to cross from one extremity of the network to the other by the shortest route. In other words, it is the shortest route between the two furthest nodes.

In our example the longest route between two nodes (in terms of the minimum number of edges one has to cross) is E to D or F to D. The minimum number of edges one has to cross to make this journey is three; thus $d = 3$. Diameter ranges from one to infinity. The higher its value, the lower the network's connectivity.

Measuring accessibility The above techniques are simply concerned with connectivity: the degree of completeness of the links between nodes. They only deal with the *number* of nodes and edges. Measures of accessibility go further than this, concerning themselves with the relative accessibility of each node to other nodes in the network. They take account of the *pattern* formed by the edges.

1 Shortest path matrix: The first stage in accessibility analysis is the construction of a shortest path matrix. To do this, letter each of the nodes in the network and then construct a grid (or matrix) based on this lettering. For each node in turn count the number of edges which have to be crossed along the shortest paths to each of the other nodes (fig. 7.17).

From: \ To:	A	B	C	D	E	F	Shimbel numbers
A	0	1	1	1	2	2	7
B	1	0	1	2	1	1	6
C	1	1	0	1	2	2	7
D	1	2	1	0	3	3	10
E	2	1	2	3	0	2	10
F	2	1	2	3	2	0	10
Gross accessibility index:							50

Fig 7.17 A shortest path matrix for figure 7.16

2 Shimbel number: The Shimbel number is an indicator of the accessibility of an individual node. It is found by adding the row of the shortest path matrix which applies to that node. Thus in the case of node A (fig. 7.16) the Shimbel number is 0 + 1 + 1 + 1 + 2 + 2 = 7. The Shimbel number of a node is simply the total number of edges needed to connect that node with all other nodes in the network individually, using the shortest paths. The *lower* the Shimbel number the more accessible the node.

3 To find an index of accessibility for the *whole network* (instead of individual nodes) use one of the following methods:
   a Add the Shimbel numbers of all the nodes. This gives the 'gross accessibility index', which is the sum of all the values in the shortest path matrix. The lower the gross accessibility index, the higher the accessibility of the network.
   b Add the Shimbel numbers for all the nodes in the network and divide by the number of nodes. This will give the 'mean Shimbel number'. In this example (fig. 7.17) the mean Shimbel number is $50 \div 6 = 8.3$. The lower the result the higher the accessibility of the network.

Connectivity and accessibility are obviously related: a network with a high level of connectivity tends to have nodes with high accessibility. But they are not exactly the same thing, as a comparison between the two networks in figure 7.18 indicates:
The connectivity measures are identical:

	a	b
n	6.0	6.0
e	7.0	7.0
$\gamma\%$	47.0	47.0

The accessibility measures, however, are different:

	a	b
mean Shimbel number	8.3	9.3
gross accessibility index	50.0	56.0

This shows that a *full* analysis of a topological network must include an assessment of both connectivity and accessibility.

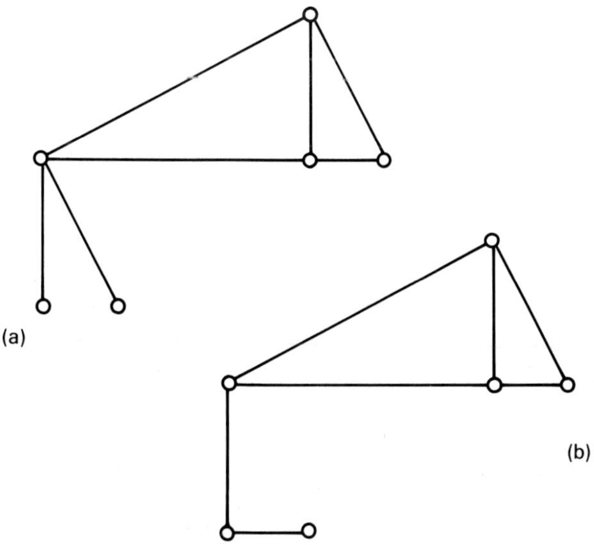

Fig 7.18 Two networks which have identical connectivity but different accessibility measures

Analysis of non-topological networks Topological transformations are carried out largely in order to simplify analysis. However, in making this simplification some of the elements of reality are lost; the network becomes less accurate. Non-topological analysis is very similar to topological analysis, the main difference being that the topological measures of connectivity cannot be used.

Measuring accessibility:

1 The shortest distance matrix: When one is given the actual distance between places it is possible to draw up a shortest path matrix using *distances* between nodes as the measures (rather than the number of edges between nodes). Two types of matrix are possible:

   a Distance from a number of places to each other (fig. 7.19).

	A	B	C
A	0	5	3
B	5	0	2
C	3	2	0

Fig 7.19 Distances (km) from three places (A, B, C) to each other

   b Distance from a number of places to other places (fig. 7.20).

	D	E	F
A	10	4	3
B	15	12	9
C	8	6	7

Fig 7.20 Distances (km) from three places to three other places

Once this has been done all the measurements discussed above in relation to the shortest path matrix of topological networks can be followed.

2 The distance Shimbel number: Add the row of the shortest distance matrix for any node to find the total distance one needs to travel to go from that node to all other nodes individually, and by the shortest route. The lower the distance Shimbel number the more accessible is the node (fig. 7.22).
3 Add the distance Shimbel numbers and divide by the number of nodes to find the 'mean distance Shimbel number', giving an index of accessibility for the whole network (fig. 7.22).

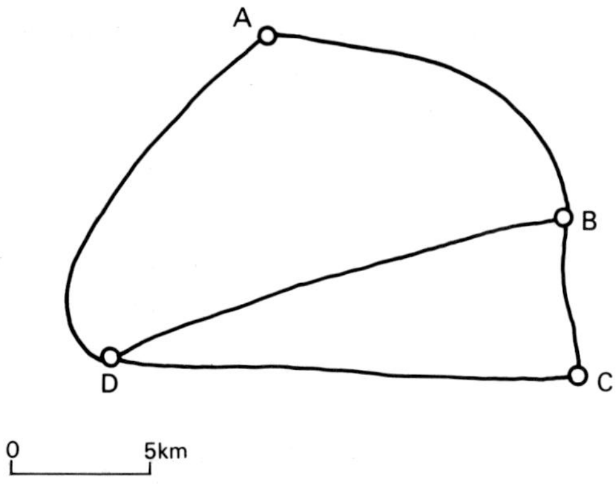

Fig 7.21 A simple real distance network

	To:	A	B	C	D	distance Shimbel number
From:	A	0	15	20	16	51
	B	15	0	5	10	30
	C	20	5	0	12	37
	D	16	10	12	0	38
						156

mean distance Shimbel no. = 156 ÷ 4 = 39

Fig 7.22 Calculation of distance Shimbel numbers and the mean distance Shimbel number, based on figure 7.21

4 Other forms of shortest path matrix: A possible alternative to the use of *distance* between places in the matrix could be the *time* it takes to travel between two nodes measured, possibly, by train journey time as given on a timetable or average non-stop car driving time. Another measure could be the *cost* of travelling between two places measured by such things as freight charges or petrol consumption. Thirdly, travel *frequency* could be used as an index; the frequency of trains between two places, for example, is easily found from a timetable.

5 Convergence count: A rather different but commonly used method for gaining a rough measure of the accessibility of a node is to count one of the following:

a The number of roads converging on the node, the number of roads of a given grade, or the number of railway lines.

b The number of trains, buses or cars converging on a node in a given time period.

Network density

There are two simple methods for summarising network density:

1 Measure the *length of route* in a given area (e.g. the number of kilometres of rail per 100 square kilometres), or in an area containing a given number of people (e.g. the number of kilometres of road per 10 000 population).

2 Count the number of *road junctions* in a given area or in an area that contains a given number of people. This method is the quicker of the two, but can only really be used for *road* networks in *developed* countries, for only under these circumstances does the number of junctions correlate well with route length.

Route sinuosity

Sometimes it may be useful to have a measure of the directness of a route between two places; that is, how far the route between two places deviates from a straight line. The basic measures are:

1 Route detour index: This expresses the actual journey distance between two places as a percentage of the straight-line distance (sometimes called the 'desire line', being the line along which people theoretically desire to travel). The formula is:

Route detour index (RDI) =

$$\frac{\text{shortest actual route distance between X and Y}}{\text{straight-line distance between X and Y}} \times 100$$

The lower the index the more direct the link. The minimum is 100 (in which case the actual distance *is* a straight-line distance); the maximum is theoretically infinity.

2 Route sinuosity: For an analysis of route sinuosity of a whole network make a matrix of route detour indices for all the pairs of nodes. Figure 7.23 gives an example which is tabulated in figure 7.24. From this matrix the node detour index can be calculated.

3 Node detour index: The node detour index is simply the average route detour index for a given node. It is obtained by adding the row associated with the node and dividing by the total number of nodes minus one. In the above example (fig. 7.24) the node detour index for A is as follows:

$$(130 + 140 + 130) \div 3 = 133.3$$

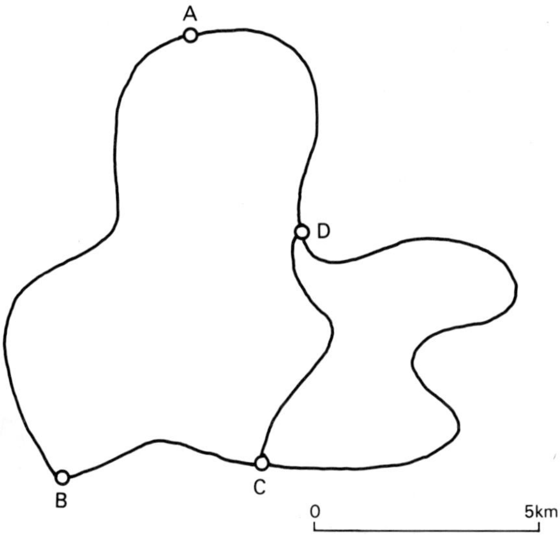

$$\text{Route detour index A–B} = \frac{13\text{ km}}{10\text{ km}} \times 100 = 130$$

Fig 7.23 A simple real distance network, with an example of a route detour index calculation

	A	B	C	D
A	–	130	140	130
B	130	–	120	140
C	140	120	–	120
D	130	140	120	–

Fig 7.24 Route detour index matrix

Traffic density

The density of traffic on a network may be measured in two ways: *either* the amount of traffic relative to the population of the area served by the network, *or* the amount of traffic relative to the length of the routes.

The amount of traffic may be expressed as either the number of vehicles using the network in a given time period, or the number of vehicle-kilometres travelled on the network in a given time period. A vehicle-kilometre is one vehicle travelling one kilometre. Two vehicle-kilometres may be one vehicle travelling two kilometres or two vehicles travelling one kilometre each.

Both these figures can be found by doing a field survey of traffic. The result may be expressed in a number of ways. Some examples are:

Number of vehicles/hour/10 000 population.

Number of vehicles/hour/100 square kilometres.

Number of vehicles/hour/kilometre of road.

Number of vehicle-kilometres/hour/10 000 population.

Ways of expressing distance and accessibility

The distance and accessibility to a place can be expressed in a number of ways:

Straight-line distance: 'from A to B is 10 km'.

Route distance: 'from A to B by road is 13 km'.

Journey time: 'it takes 12 minutes to go from A to B'.

Journey costs: 'it costs 41p to go from A to B'.

Perceived distance: 'I *think* it is 15 km from A to B'.

Journey effort: 'it takes 39 units of energy to move from A to B'.

Flows *possible* along routes: '34 cars per minute can pass along the road from A to B'.

You may be able to think of other ways of expressing distance and/or accessibility. When we talk about one place's accessibility to another place in terms other than straight-line distance, we are talking about their position in *relative* space. Relative space is often a more realistic measure of accessibility than actual physical space.

The measures given above can be plotted cartographically as isolines (page 78); lines of equal travel cost are called isotims, whereas lines of equal travel time are called isochrones. By drawing circular isochrones at regular intervals around one particular place (A) and plotting other places in relative space according to travel time from A, one may obtain a visual impression of the spatial maldistribution of a transport network (fig. 7.25).

7.6 Exercises

1 Figure 7.26 shows the location of publishers, printers and newspaper offices in the City of London. Find the median centre of this distribution (page 105).

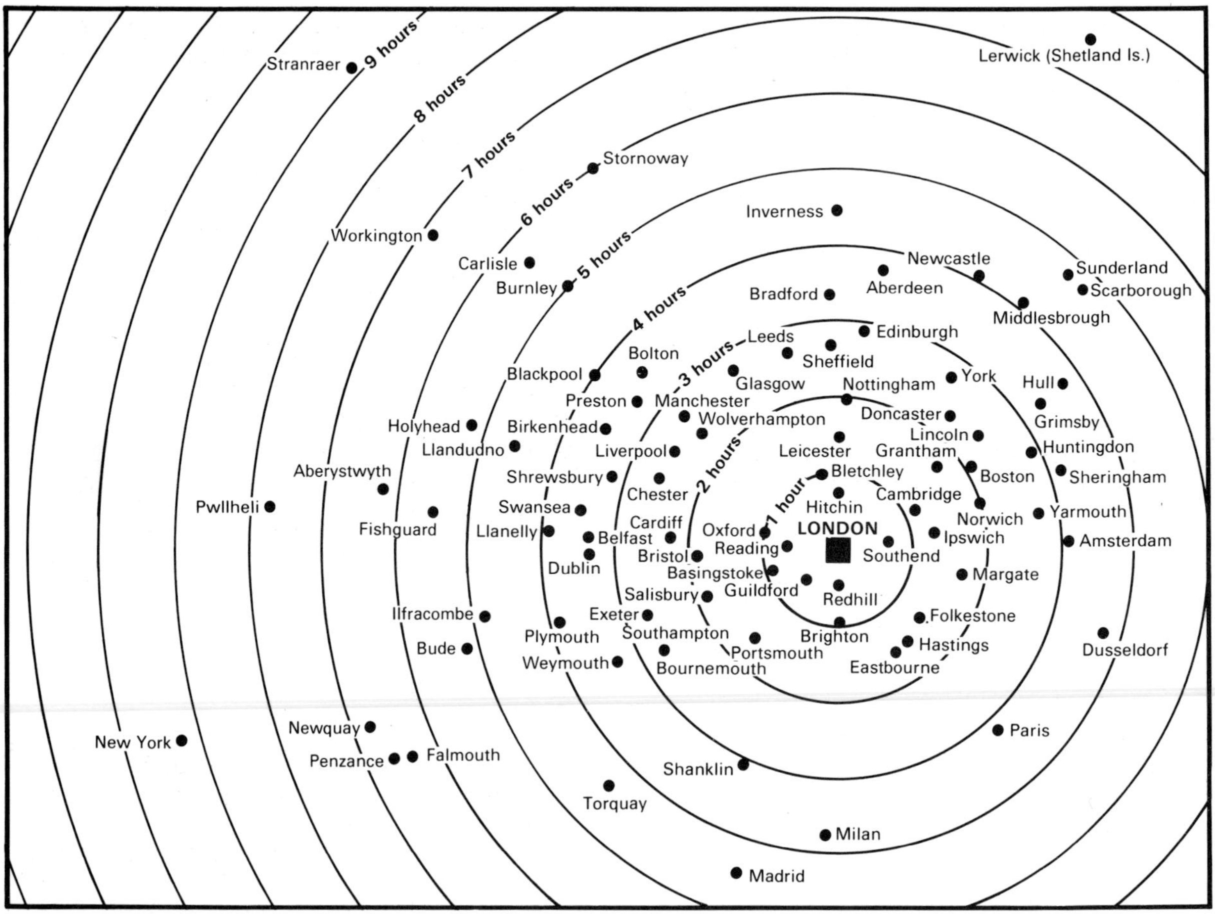

Fig 7.25 Map showing isochrones of travel times, by the fastest generally available means, from London to parts of Britain, the Continent and North America

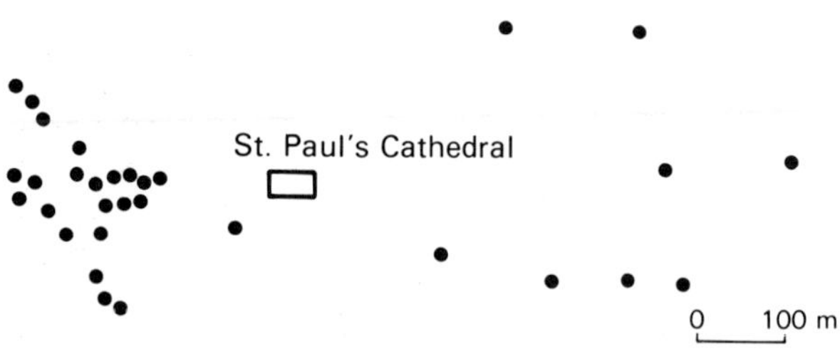

Fig 7.26 The distribution of publishers, printers and newspapers in the City of London in 1980

2 Figure 7.27 shows the location of restaurants in Torquay, Devon. Calculate the nearest neighbour index for this distribution (page 105).

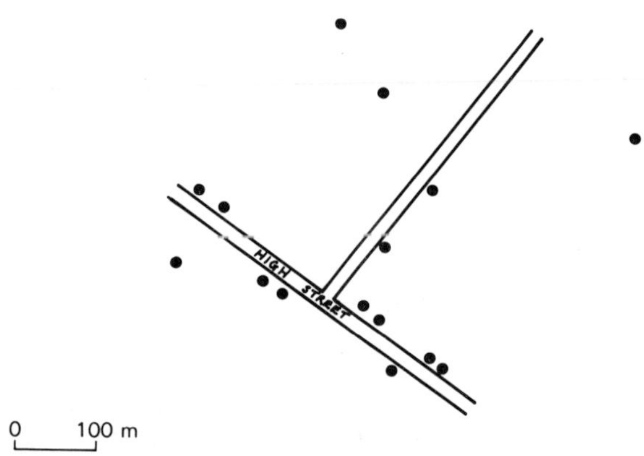

Fig 7.27 The distribution of restaurants in Torquay

Sources and further reading

General texts:

Boyce, J., and Ferretti, J., 1984, *Fieldwork in Geography,* Cambridge University Press.

Daugherty, R., 1974, *Data Collection*, Oxford University Press.

Davis, P., 1974, *Data Description and Presentation*, Oxford University Press.

Hammond, R., and McCullagh, P., 1978 (2nd ed.), *Quantitative Techniques in Geography*, Oxford University Press.

Matthews, M. H., and Foster, I. D. L., 1986, *Fieldwork Exercises in Human and Physical Geography*, Edward Arnold.

Matthews, M. H., and Foster, I. D. L., 1989, *Geographical Data: Sources, Presentation and Analysis*, Oxford University Press.

McCullagh, P., 1974, *Data Use and Interpretation*, Oxford University Press.

Monkhouse, F. J., and Wilkinson, H. R., 1971 (3rd ed.), *Maps and Diagrams,* Methuen.

Pilbeam, A., 1980, *Local Projects in A-Level Geography*, Unwin Hyman.

Human geography:

Morgan, M. A., 1979, *Historical Sources in Geography*, Butterworths.

Short, J. R., 1980, *Urban Data Sources*, Butterworths.

Toyne, P., and Newby, P. T., 1971, *Techniques in Human Geography*, Macmillan.

Physical geography:

Briggs, D. J., 1977, *Sediments*, Butterworths.

Briggs, D. J., 1977, *Soils,*, Butterworths.

Finlayson, B., and Statham, I., 1980, *Hillslope Analysis*, Butterworths.

Gilbertson, D. D., Kent, M., and Pyatt, F. B., 1985, *Practical Ecology for Geography and Biology*, Hutchinson.

Goudie, A. S., (Ed.), 1981, *Geomorphological Techniques*, Unwin Hyman.

Goudie, A., and Gardner, R., 1985, *Discovering Landscapes in England and Wales*, Unwin Hyman.

Gregory, K. J., and Walling, D. E., 1973, *Drainage Basin Form and Process,* Edward Arnold.

Morisawa, M., 1968, *Streams*, McGraw-Hill.

Oke, T. R., 1987 (2nd ed.) *Boundary Layer Climates*, Methuen.

Williams, G., 1987, *Techniques and Fieldwork in Ecology*, Unwin Hyman.

Acknowledgements

The authors and publisher are grateful to the following individuals and organisations for the use of data and/or ideas:
fig. **3.1**, Rahns P, 'The weathering of tombstones and its relationship to the topography of New England', *Journal of Geological Education*; figs. **3.9**, **3.27**, Ingle Smith D, and Stopp P, *The River Basin*, Cambridge University Press; fig. **3.34**, Glavin C J, (1968) 'Breaker type classification on three laboratory beaches', *Journal of Geophysical Research 73*, 3651–9; fig. **3.40**, Briggs D, *Sediments*, Butterworth & Co. Ltd.; fig. **3.44**, Hilton K, *Process & Pattern in Physical Geography*, University Tutorial Press Ltd.; figs. **3.52**, **3.54**, **3.55**, **3.56**, **3.58**, **3.59**, **3.60**, **3.66**, Fitzgerald B P, *Weather in Action*, Methuen Educational Ltd.; fig. **3.53**, Gates E S, *Meteorology and Climatology*, Harrap; fig. **3.65**, taken from Hilton K, *Process and Pattern in Physical Geography*, University Tutorial Press Ltd.; fig. **3.68**, Oxford & Cambridge Schools Examination Board, 'A' level Geography Paper 1, question 8, June 1980; **3.70(a)**, Satellite Receiving Station, Department of Electrical Engineering, University of Dundee; fig. **4.5** The Controller, HMSO and Chas. E Goad Ltd.; fig. **4.14**, Kelly's *Tradefinder* (photographer: George Williams); fig. **5.8**, Clarke C G, 'Residential Segregation and Intermarriage in San Fernando, Trinidad', *The Geographical Review*, Vol. 26, No. 2; figs. **5.12**, **5.13**, **5.14** Vaughan-Williams P, *Brasil: A Concise Thematic Geography*, University Tutorial Press Ltd. (sources: A.N.D.B., plate iv–6 and EMPLASA: Sumario de dados da Grande São Paulo, 1977); fig. **5.15**, Office of Population Censuses and Surveys, *Census 1981: Preliminary Report*, (code n. CEN PR (1), HMSO; fig **5.16**, Daniel P, and Hopkinson H, *Geography of Settlement*, Oliver & Boyd Ltd.; fig. **5.18**, Berry, Brian J L, *Geography of Market Centers and Retail Distribution*, © 1967, pp. 11, 12. Reprinted by permission of Prentice-Hall, Inc., Englewood Cliffs, N.J.; fig **5.24**, Coppock J T, 'The cartographic representation of British agricultural statistics', *Geography*, Vol 40, pt 2, April 1965; fig. **6.1** Gregory S, *Statistical Methods & the Geographer* (second edition 2), Longman Group Ltd.; fig. **7.10**, London Transport Executive; fig. **7.25** adapted from fig. 5.32 (p. 194) in *Location in Space: A Theoretical Approach To Economic Geography*, second edition by Lloyd P E, and Dicken P. Reprinted by permission of Harper & Row, Publishers, Inc.; appendices **2 & 3**, Lindley D V, and Miller J C P, *Cambridge Elementary Tables*, Cambridge University Press; appendix **4**, Fisher R A, and Yates F, *Statistical Tables for Biological, Agricultural and Medical Research* (fig. **15**), published by Longman Group Ltd., London, (previously published by Oliver & Boyd Ltd., Edinburgh) and by permission of the authors and publishers; appendix **5**, Tidswell V, *Pattern & Process in Human Geography*, University Tutorial Press Ltd.; (source: McCullagh P, *Data Use & Interpretation*); p. **3**, Oxford & Cambridge Schools Examination Board, part of a list of titles of successful fieldwork essays circulated by the Board to its schools in 1979: pp. **7**, **8**, **91**, **98**, **121**, Huff D, *How to Lie with Statistics*, Penguin Books Ltd., cartoons by Mel Calman (photographer: George Williams).

The authors and publisher would like to thank the Literary Executor of the late Sir Ronald A Fisher, F.R.S., Dr. Frank Yates, F.R.S. and Longman Group Ltd., London for permission to reprint Table 15 from their book *Statistical Tables for Biological, Agricultural and Medical Research*. (6th edition, 1974).

Fig. **4.2** was reproduced with the permission of Ronald Johnson, Vicar of Dorney. All photographs, with the exception of those cited above, were taken by Paul Cleves.